The Globalisation of High Technology Production

THE GLOBALISATION OF HIGH TECHNOLOGY PRODUCTION

Society, space and semiconductors in the restructuring of the modern world

JEFFREY HENDERSON

London and New York

First published 1989 by
Routledge
11 New Fetter Lane, London EC4P 4EE

Simultaneously published in the USA and Canada
by Routledge
a division of Routledge, Chapman and Hall, Inc.
29 West 35th Street, New York, NY 10001

Reprinted 1990

New in paperback 1991

Typeset in Times by Witwell Ltd, Southport
Printed and bound in Great Britain by
Billings & Sons Limited, Worcester

British Library Cataloguing in Publication Data

Henderson, Jeff, *1947–*
The globalisation of high technology
production: society, space and semiconductors in the restructuring of the
modern world.
1. High technology. Socioeconomic aspects
I. Title
306'.3

ISBN 0–415–06076–1

Library of Congress Cataloguing in Publication Data

Henderson, J. W. (Jeffrey William), 1947–
The globalisation of high technology production: society, space, and semiconductors in the restructuring of the modern world/
Jeffrey Henderson.

p. cm.
Bibliography: p.
Includes index.
ISBN 0–415–06076–1 (Pbk)
1. Semiconductor industry. 2. Electronic industries.
3. International division of labor. 4. Offshore assembly industry.
I. Title.
HD9696.S42H46 1989 88–32173
338.4'762138152—dc19 CIP

To my mother and brother, and to the memory of my father

Contents

List of figures and tables

Figures

Tables

Foreword

Looking backwards, it now seems ironic that the theoretical model which dominated the literature on the sociology of development for over a decade was actually about the impossibility of development in the Third World. The title of one of A. G. Frank's highly influential books on 'dependency theory', for instance, postulated as the only alternatives 'Underdevelopment or Revolution'.

In the epoch of Ché Guevara and successive wars in Viet Nam, the second part of this model was more than rhetoric. But there have been few revolutions since then, and the largest of them, in China, has now taken a quite different course, while smaller revolutionary regimes – Angola, Mozambique, and Nicaragua – have been subjected to brutal destabilisation.

The conditions that led to those sustained and bloody struggles still exist in many countries. Yet the image which the West has of the Third World today is no longer one of a zone of potential peasant revolution. It is, rather, one of irrational nationalism, of religious fundamentalism, and of ethnic conflict, all of them fed by chronic and inescapable rural poverty, by technological backwardness and stagnation, and by helplessness and apathy in the face of Superpower, especially First World domination. The constant efforts of Third World peoples to solve their own problems – whether at the level of challenges to political overlords and to entire social systems, or at the much less dramatic level of the localized activities of non-governmental organisations – are often obscured by the assumption that help necessarily has to come from the outside, from the hands of relief workers and experts, or in the form of foreign capital and know-how.

The basic image that results is that the entire Third World is a disaster zone, a notion uncritically accepted by millions who receive their information about the Third World primarily from organisations dedicated to relieving famines, floods, and other catastrophic forms of 'natural' disaster (actually often the results of man-made policies, from cash-crop farming to destruction of the forests). For most people in the West, therefore, the dominant image of the Third World – diffused by organisations such as Oxfam and War On Want, and stamped on to the consciousness of millions by sustained appeals to their charity – is that of a skeletal African child with his hands outstretched for food.

The growth of industry, and the concomitant explosion of cities, generally only penetrates public consciousness in so far as it is known to have sucked in millions of peasants to the shanty towns of São Paulo, Calcutta, or Nairobi. That they went there because they were attracted by the reality of rapidly expanding industry and commerce, which has provided employment for a growing new working class, often goes unrecognised or unacknowledged.

People who actually visited the Third World, of course, soon found that it was no longer a sea of peasants and that the 'marginals' in the shanty towns were not the only inhabitants of Third World cities, not even the majority. Both popular stereotypes and the theoretical orthodoxies and fashions, which have successively dominated development studies, were long overdue for reappraisal.

The most systematic attempt to propound a more viable theoretical framework, Immanuel Wallerstein's 'world-system theory', began by reasserting the central Frankian theme: that particular national cases of development and underdevelopment could only be understood by locating them within a model of the world-system as a whole.

But by introducing a new category between 'core' and 'periphery', the 'semi-periphery', the brute fact that development was actually taking place in many Third World countries was given theoretical recognition. Though it remained an uneasy, residual category, and ill-defined, it did mitigate somewhat the determinist cast of a model of the world-system polarised into industrialised and agrarian segments, the latter so much under the control of those at the 'centre' that it was difficult to see how any particular country could ever escape from control exercised from the centres.

Following traditional Trotskyist analysis, moreover, the major real-world alternative to capitalism – one of the now many kinds of communist society – is depicted as no alternative at all, since all are dependent variables within a unitary and capitalist-dominated world-market, if not simply statal forms of capitalist production.

Only a few countries have taken that road out of dependency. But an increasing number have been able to escape from underdevelopment precisely along the route which Frank excluded as a possibility: via capitalist transformation of the economy. More recent theoretical attempts to create a better model of Third World trends have therefore had to become more developmental.

Bill Warren's *Imperialism: Pioneer of Capitalism* (Warren 1980) did so in a singularly Eurocentric and optimistic (not to say Victorian and unilinear-evolutionist) way: by a provocative updating of Marx's analysis of the 'progressive' implications of the spread of capitalism across the globe, in the form of nineteenth-century colonialism – not only in respect of the transformation of the forces of production (the introduction of plantations, cash crops, mines, and factories), but also in terms of concomitant improvements in the quality of life (education, health, social services, etc. etc.) – an image of capitalism naturally unacceptable to more orthodox Marxists was generated. Representatives of this group, such as Nigel Harris, confined themselves mainly to the much less benign diffusion of capitalist technology and production-relationships. Harris's model was also evolutionist, but in a different way. It was a vision of a process seemingly so inexorable that though it had so far only occurred in a handful of countries – the transformation of Brazil, Mexico, and the Asian 'Gang of Four' (South Korea, Singapore, Hong Kong, and Taiwan) from 'newly-industrialising' countries to actually industrialised ones – it was nevertheless 'the wave of the future' everywhere; indeed, it presaged the imminent *End of the Third World* (Harris 1987) as a whole. Yet for the inhabitants of most of an entire continent, Black Africa, for instance – where GDP has actually declined over the last decade – such an over-generalised model, based on a set of exceptional instances which include two highly specialised and anomalous city-states (one of which flourishes in part because it is the doorway into and out of China) is as lacking in credibility, as a general picture of what is happening to the underdeveloped world, as those models which continue to present the Third World as a zone of rural squalor. Hence countries like Ecuador or Sri Lanka are not likely to be persuaded that their future is to become mini-Brazils.

Third World countries share a common economic market-situation *vis-à-vis* the developed world, to the extent that they remain victims of their common colonial past, despite political independence, as producers of primary commodities on a world-market in which manufactured goods command higher prices, and where ownership and control of enterprises in their countries are in foreign hands.

But they are also divided in important ways, even in economic terms, because there are major differences in the ways in which the different economic sectors within their countries are inserted into the global division of labour: some are oil-producers; others are producers of agricultural commodities, whilst yet others have become parts of a sophisticated high-tech industrial division of labour. And others are only significant in world terms because of their strategic, not their economic utility.

Somewhat surprisingly, given its intellectual line of descent, some theoreticians of the new international division of labour have neglected these modern forms of the phenomenon which Trotsky insisted, was typical of modern capitalism: uneven development. Instead, their analysis has rested, far too much, upon a selective use of predominantly macro-economic statistical data relating to the entire Third World. Ignored were the enormous differences between their economies, and differences in the relationships between the various sectors and the industrialised world, not to mention the extraordinary variety of social institutions and of cultures in more than a hundred countries subsumed under the single label 'underdeveloped'.

Nor did the intensive and qualitative study of particular industries, industrial sectors, or regions appeal to those who were basically concerned to repeat the proposition that the characterisation of the whole, the world-system, had to precede any attempt to analyse any of its parts: case-studies, say, of particular industries, industrial sectors, or regions.

It is also unfortunately true that case-studies of industrial development in the Third World have not only been few and far between, but also poor in quality. One outstanding feature of Jeffrey Henderson's study of the semiconductor industry is that he does not relapse into an 'empiricist epistemology'. His starting-point, rather, is the emergence and the continuing reality of a global system of production (and not merely of exchange). But he has devoted many years to the careful study of one particular and highly strategic industry in a world in which communications constitute a leading edge of development in all fields: the semiconductor industry, firstly in its heartland, Silicon Valley, California, and subsequently in Southeast Asia, Hong Kong and in Britain (particularly Scotland).

He therefore found it necessary to ask himself a question that over-generalised world-system theory rarely addresses: what it was that induced foreign corporations to locate their offshore plants not just in an undifferentiated 'Third World', but in some particular part of it: in Hong Kong, for instance, rather than elsewhere. The relevant factors include not only directly industrial policies which made these countries financially attractive to potential foreign corporate investors – tax conditions, the repatriation of profits, trade union legislation, etc. – but also the existence of key social preconditions: the availability both of low-level (but adequately educated) labour and of higher-level scientists and engineers; of back-up resources in the shape of scientific, technological, and scientific institutions; and the existence of a stable labour force as a result not simply of repression, but of a variety of social policies ranging from the provision of housing to the control of the hours of work.

Some of these conditions (wage-levels, work-conditions, promotion prospects, pension rights, etc.) are predominantly under the control of the corporations themselves, but many have evolved as a result of governmental initiative to attract investment long before it actually arrived. Government policy remains, therefore, an independent, local variable.

Hence un-dialectical early models of dependency, which assumed that all the key decisions were taken by the owners of corporate capital in the 'centre', now look inadequate. They also look ahistoric and fatalistic, because they assumed that the early phases of the new international division of labour – when capital moved to low-wage countries employing predominantly female labour, primarily in the textile and other industries which were being displaced by new high-tech industries in the West – would continue to be the permanent and typical features of centre–periphery relationships.

Henderson's book covers an immense enough canvas. It will be up to others to build upon this pioneering study: to study, for instance the more subjective and socio-cultural aspects of the new industrial society: the apoliticism of an immigrant, insecure population; its 'familistic instrumentalism'; its pragmatism; the weakness of trade unionism; its readiness to accept limitations upon its capacity to express itself politically and to develop its own institutions (not just trade unions, but social associations of all kinds), in return for security and rising material living conditions: issues which have been suggestively discussed in *Society and Politics in Hong Kong* (Lau 1982).

True, in country after country: in Taiwan, in South Korea, in Brazil, and elsewhere: rising living standards have not been any substitute for a growing demand for the right to participate in the decision-making processes that affect the lives of everyone in all spheres of life. In East Asian societies which Jon Halliday described only a few years ago as 'semi-militarized', the demand for democracy is no longer treated, even by hard-line radicals, as a 'mere' bourgeois demand: it has moved to the centre stage of politics. But it has been the new intelligentsia which the new industrialism has brought into being, the students who have raised the banner of democracy, not the class which classical Marxism identified as the one which would lead the struggle for human emancipation: the working class.

I have strayed beyond Jeffrey Henderson's already wide-ranging brief. That is because it is the kind of book that opens horizons and makes one ask questions about where society is going: not just in East Asia, not even in the Third World alone, but globally. There is much here that will particularly interest those who are concerned with the 'centre', and with the conversion of historic heartlands of industry like Scotland into the semi-peripheries of an industry controlled from the USA and Japan.

This is a creative and readable pioneer study; a model of sociology at its best. I am sure that it will be seen, one day, as a classic.

Peter Worsley

Preface

All too often in the past our ability to comprehend the processes of economic development, its dynamics and consequences, has been restricted by the relative absence of studies which detailed the operation of particular industrial sectors, branches, or firms, as they emerged, expanded and perhaps declined in the context of particular societies. Students of what are now advanced industrial societies, however, were always more 'privileged' than others in the sense that they had a plethora of industrial and labour histories and a substantial body of work by sociologists and economic geographers on which to draw. Those who studied economic change in the 'developing' world, for a variety of reasons were unable to benefit from such advantages. They tended to work largely at a macro level, struggling with aggregate data, bolstered where possible by the ethnographic studies of anthropologists and sociologists.

In the last few years this situation has begun to improve. Social scientists of development have begun to focus on the processes of change within and between the components of the economies and societies of the 'Third World'. In some cases they have done this while keeping a theoretical brief for the dynamics of development at the global level. The work presented in this book is intended, in part, to be a contribution to this growing body of literature. Hopefully, however, whatever merits it may have, also lie elsewhere. As well as investigating the growth of an important industry in one region of the less-developed world, it also, at the same time, tries to assess its significance for one of the societies of the more-developed world. With luck, the account of the industry's development and consequences in all of these societies has been given a certain degree of unity, and rendered relatively coherent, by being cast in terms of an explicitly global theoretical framework.

Specifically, this study seeks to analyse the semiconductor industries of the developing countries of the East Asian region, as well as Scotland, in relation to the international division of labour of which they are a part. It does this by theorising the emergence and subsequent development of the industry in these locations, and in terms of their structural connections with American semiconductor production in the USA. In this context, the study explores the social and spatial dynamics and implications of the industry's development for the various territorial units in which it has taken root. It suggests that semiconductor and similar forms of 'high technology' production constitute a new mode of industrialisation which may have different implications for economic and social development than had earlier modes. This proposition is investigated initially in the industry's principal 'home base', Santa Clara County, California, but especially in East Asia and Scotland. The study shows that a distinct regional division of labour is emerging in East Asia (with its own 'cores' and 'peripheries') which could not have been 'predicted' from recent interventions in development theory. This division of labour has different implications for the particular territorial units incorporated within it. Additionally, while Scotland has a more dominant and autonomous role within the semiconductor industry's global division of labour, than practically all of the East Asian locations, it has serious problems which may affect its ability to assist the rejuvenation of the Scottish economy. The book continues with an assessment of the prospects for future development in each location against the backdrop of the continuing restructuring of the industry at the global level. It concludes with a number of comments on the significance of the semiconductor industry in East Asia and Scotland for analyses informed by the international division of labour perspective.

The research this book discusses was conceived one warm May morning in 1981. In conversation with Robin Cohen, and lazing on one of the lawns of the Warwick campus, the idea of investigating the changing relation of Britain and Hong Kong to the international division of labour was born. Thanks to the higher-education policies of the Thatcher government, I was about to be exiled to Britain's last colony of any significance. My move, though at first unwelcome, seemed to provide us with a valuable opportunity to study the dynamic interconnections between deindustrialisation in Britain and industrialisation in Hong Kong. Though Robin began to work on Britain, other commitments, particularly after he became Director of Warwick's Centre for Research in Ethnic Relations, ensured his early retirement from formal participation in the project. Only one paper remains to attest to this particular collaboration (Henderson and Cohen 1982a).

With my becoming solely responsible for the project, the focus of the research began to change. Empirically it was both narrowed and expanded in the sense that it began to concentrate specifically on semiconductor production, but on a broader geographical canvas. Theoretically it was expanded by taking on board more of a spatial cast than originally had been intended. In spite of these changes, Robin Cohen maintained his commitment to the enterprise, continuing to provide inspiration, encouragement, and critical comment. The fact that I am able to write these lines, at this time, owes a great deal to his support, not merely in the last seven years, but across nearly two decades of friendship.

My other major debt is to Manuel Castells. Ever since our first meeting in Mexico City in August 1982, Manuel has been a constant source of encouragement. Through his own work (which in recent years has become concerned, in part, with the developmental consequences of high technology industrialisation), his comments on my work and his support and friendship (renewed since our first meeting in places as disparate as Hong Kong, Madrid, Berkeley and Delhi), he has played a major role in helping me to believe in my abilities as a scholar. In particular, he helped develop my confidence in this study as a worthwhile enterprise.

In addition to Robin and Manuel, two other scholars and friends have been particularly helpful. Allen Scott helped to reorient my thinking about the research at a critical moment in its development, and he collaborated with me on a related paper, parts of which are reproduced in Chapter three of this book. Richard Child Hill was kind enough to read the entire manuscript when it was in an advanced, though still draft form, and he was able to alert me to a number of its remaining deficiencies.

Over the period of research a number of individuals and organisations have provided considerable help. Tang Wang Shing, James Kung, Christine Chau, Eric Kwok and Chan Yin Sang provided various forms of assistance on this and related work. Lenny Siegel gave me access to the Pacific States Centre's extensive files on the US electronics industry and his own detailed knowledge of Silicon Valley. David Donnison, Bill Lever, Bob Miles, Ivan Turok, John Eldridge, Peter Cressey, John MacInnes, Jack Parr and Michael Danson provided advice and succour from their base at the University of Glasgow during a research visit to Scotland. David Wellman and Troy Duster provided me with a base at the University of California at Berkeley's Institute for the Study of Social Change, while I did research in California. The University of Warwick, and particularly its Department of Sociology, provided me with an Associate Fellowship, 1981–5, and a congenial intellectual home when I was in Britain. Jonathan Schiffer, Daniel Han, Jim Newton, Murray Groves, Yao-Su Hu and Harry Dimitriou, colleagues both past and present at the University of Hong Kong, improved the intellectual atmosphere and provided crucial emotional support in periods of declining morale. Worthy of special mention is my colleague Sandy Cuthbert, who from the beginning of my sojourn in Hong Kong has been my soul brother in the struggle for academic and personal integrity, often in the face of overwhelming odds. Additionally, without Belinda Man's speedy and efficient typing of numerous drafts of various sections and chapters, this book would not have seen the light of day for some time to come.

Funding for various aspects of the research was provided by the Lipman Trust, London, the University of Hong Kong's Urban Studies and Urban Planning Trust Fund, and the Board of Management of *Urban Studies* (in the form of appointment to the Urban Studies Fellowship for 1985). In addition, the University of Hong Kong provided me with the leave necessary to undertake research in Scotland and with a grant which partially funded my research in California. The University of Melbourne awarded me a Visiting Fellowship in late 1985. It was while I was a temporary member of its School of Environmental Planning, that I was able to draft parts of the paper that ultimately became Chapter six of this book. Finally I am grateful to the managers of various semiconductor companies throughout the world for allowing me to interview them and for providing information. I owe a great debt to all these individuals and organisations for their generosity, help, and support.

In addition to the friends and colleagues cited above, a number of others have provided comment and criticism on earlier versions of particular chapters. David Slater commented on the substance of Chapter two. Doreen Massey, Chris Pickvance, and Lenny Siegel provided critical feedback on Chapter three, while Jonathan Schiffer commented on both of these chapters. Stewart Clegg, Phil Cooke, Kevin Cox, Peter Hall, and Andrew Sayer commented on Chapter six. During the final stages of the book's preparation, Ivan Szelenyi and Peter Worsley expressed particular interest and support. While I have not taken on board all of the suggestions of my commentators and critics, without doubt their efforts have helped to improve the quality of my work, and for this, I am deeply grateful. Whatever problems remain, are, of course, entirely my responsibility.

My final debt of gratitude is to Ho Shuet Ying. She was responsible for assisting me with part of the research, but more importantly she helped me to experience the good times and pulled me through the bad times that are an inevitable by-product of all forms of wage labour, including its intellectual varieties.

Jeff Henderson
Hong Kong
July 1988

Acknowledgements

Earlier versions of parts of this book have been published elsewhere. The author, therefore, would like to thank the copyright holders below for their permission to utilise that work:

David Drakakis-Smith for 'The New International Division of Labour and Urban Development in the Contemporary World System' which appeared in D. Drakakis-Smith (ed.) *Urbanisation in the Developing World*, London: Croom Helm, 1986; Michael Breheny and Ronald McQuaid for 'The Growth and Internationalisation of the American Semiconductor Industry: Labour Processes and the Changing Spatial Organisation of Production' which appeared in M. Breheny and R. McQuaid (eds) *The Development of High Technology Industries: An International Survey*, London: Croom Helm, 1987, (co-authored with A. J. Scott); Chris Dixon, David Drakakis-Smith and H. D. Watts for 'The New International Division of Labour and American Semiconductor Production in Southeast Asia' which appeared in C. Dixon, D. Drakakis-Smith and H. Watts (eds) *Multinational Corporations and the Third World*, London: Croom Helm, 1986; The Board of Management of *Urban Studies* for 'Semiconductors, Scotland and the International Division of Labour', which was published in *Urban Studies*, 24(5), October 1987.

Chapter one

A new mode of industrialisation

The world economy, for the past twenty years or so, has been subject to persistent and ever-deepening crisis tendencies. At the level of the firm, the response to this situation has been the initiation of major processes of organisational and technical restructuring. Now, as in previous periods of crisis, restructuring is predicated on the search for new bases for capital accumulation. Today, however, with the increasing integration of the various units of the world economy, and the growing dominance of the transnational corporation, restructuring operates at the global as well as the national scale. As a result, current restructuring has begun to have far-reaching economic, social, political, and spatial implications for both 'developed' and 'developing' societies and for the connections between them (Henderson and Castells 1987). Although there is now a significant literature which examines the contours and impact of restructuring in particular territorial units, be those units countries, regions or cities (cf. Blackaby 1979; Bluestone and Harrison 1982; Massey and Meegan 1982; Massey 1984; Newby, *et al.* 1985), there is so far relatively little work that explicitly examines restructuring in given spatial locations in relation to the dynamics of the changing world economy (see Armstrong and McGee 1985; Walton 1985a; Henderson and Castells 1987, for exceptions). Furthermore, while there is some work which charts global restructuring across a number of industries (cf. Dicken 1986), there is little of it which examines how the dynamics of global economic restructuring produce different, but related effects in globally disparate territorial units.

This study is an attempt to fill these gaps. It is inevitably, however, a limited attempt. Global restructuring operates through and across all circuits of capital (Harvey 1982, 1985; O'Connor, 1984) and all circuits (productive, commodity, money) are composed of separate sectors, branches, and firms organised nationally or transnationally, and subject to different sets of determinants (associated with the nature of markets, technology, the historical and cultural traditions of the territorial units they inhabit etc.)[1] To address global restructuring in its totality, and take account of its dynamics and consequences for various parts of the world, would require, therefore, a series of related research agendas with armies of researchers operating in a variety of countries, regions, and cities. While there are signs that such international networks of scholars with relevant agendas are beginning to emerge,[2] there are serious constraints on what one scholar, with limited time and resources, can achieve on his or her own.

This study, then, arises out of concerns to understand the relation between economic, social, and spatial change in particular territorial units and the dynamics of global restructuring. It pursues its investigation with particular regard to industrial development in both a (perhaps former) core society, as well as in a number of the developing societies of East Asia.[3] Its specific empirical focus, however, is not the industrial economies of those societies in their entirety, nor even one of their sectors. Rather, it concentrates on one important branch of both these and the world industrial economy generally: semiconductor production. While much of what follows in this introductory chapter will attempt to provide a rationale for this focus, it is first necessary to indicate those elements of contemporary restructuring that I take to be historically unique.

The global option

Efforts to deal with current crisis tendencies have relied, in part, on forms of restructuring that have been evident as long as there has been industrial capitalism. So for instance, social relations of production have been altered by means of a reorganisation of labour processes designed to ensure increases in the pace of work and in supervisory control. This, together with reductions in excess productive capacity, has resulted in growing and longer-term unemployment in many core economies (on the British case, see, for instance, Massey and Meegan 1982). These developments, combined with an increased belligerence of some governments towards organised labour, have led to a weakened trade union capacity to control wages and working conditions. As in earlier crises, new technologies (but this time based on microelectronics in the form of CAD/CAM systems, etc.) also have been utilized as part of the restructuring process. Furthermore, a process of capital concentration has taken place associated with the elimination (or incorporation) of weaker competitors.

In previous crises, restructuring was usually confined to the given territorial unit. What is new about the contemporary period, however, is that restructuring 'internal' to the territorial unit has been combined with spatial (both intra- and inter-national) shifts in investment and a massive expansion of the radii of organisational control associated with the growth of transnational corporations.[4] It is precisely this combination of the increased use of space together with the expanded transnationalisation of corporate structures that has given the current restructuring process its global character and dynamic.

To stave-off cut-throat competition and generate new accumulation possibilities, core capital, then, has increasingly turned to this 'global option'. The emergence of the global option, however, would have been inconceivable without the development of information technologies, and particularly telecommunications. These technologies have been a major material condition for the emergence of the global option in as far as they have enabled particular labour processes, or sometimes entire production facilities, to be dispersed across the globe, while allowing managerial control (and in the case of industrial production, often the most advanced design functions) to remain centralized in the 'world cities' of the core societies (Friedmann 1986). So central have been these new microelectronic technologies to the recent development of the international economy, that elsewhere Castells and I have suggested that global restructuring, at root, must be considered as a 'techno-economic' process (Castells and Henderson 1987).

A new mode of industrialisation

However important telecommunications might be to the implementation of the global option, they are but one form of the microelectronic (or high technology) 'revolution' of recent years. The varied products which constitute the basis of that 'revolution', however, have one thing in common. They have the same technological core, and that core is the semiconductor in the form of the transistor or diode, but particularly the integrated circuit ('silicon chip'). Though semiconductors do not rank among the world's leading industrial products in terms of the value of production, they arguably constitute the most important industrial branch of the contemporary epoch. Not only are they the central component in the transmission, reception and amplification of electronic signals and as such essential to telecommunications, data storage, retrieval, and manipulation (computers), but they are also basic to home entertainment (TV, video, hi-fi etc.), medical science, military hardware, aerospace etc. Without semiconductors, industrial societies would be unworkable, and social life within them, almost unthinkable.

In addition to their central role in the modern world, high technology industries, but perhaps semiconductors in particular, have been seen by economic analysts and governments as one of the principal solutions to the problems of economic crisis and national development. These beliefs have been spurred on and given popular credence by the media-generated imagery associated with the world's most important high technology location, Santa Clara County ('Silicon Valley'), California. For all these reasons, and more, 'semiconductors are now accorded a strategic role for national economies similar to that of steel fifty years ago' (Morgan and Sayer 1983: 18).

The influence of high technology products on human existence and the strategic role semiconductors in particular play in (some) national economies are insufficient in themselves to warrant my depiction of industrial development, based on microelectronics, as a 'new mode of industrialisation'. What renders this depiction sufficient, however, is my contention that industrialisation (or reindustrialisation) based on new technologies takes a different form, and may well have different consequences from previous rounds of industrialisation, based as they were on steel, engineering, automobiles, shipbuilding, textiles, and the like. Bearing in mind that semiconductors are not only the heart of microelectronics and information industries generally, but that semiconductor companies themselves constitute a production and organisational form that is a paradigm example of the global option in practice (see Chapters 3–6), we can begin to specify what it is about the manufacture of semiconductors and electronic products more generally, that may have brought into being this new mode of industrialisation.

There seem to be four elements which distinguish electronics industries from those manufacturing processes which formed the basis of the initial industrialisation of the core economies. These elements, therefore, are the defining characteristics of the new mode of industrialisation:

1. Electronics industries utilise a distinctive raw material: knowledge. They have relied not only on highly creative scientific and engineering expertise for their initial development (that much has been true historically for most manufacturing industries), but also they have depended on continuing technological innovation as perhaps the primary basis of their competitive 'edge'. Arguably then, knowledge, embodied in a particular form of labour power, has been the principal factor of production. That form of labour power, however, is not evenly distributed within territorial units, nor across the globe (only unskilled labour power is), and this, in concert with other factors, as we shall see (in Chapters 3–7) has

had significant consequences for the spatial distribution of production facilities.

2. A fundamental task of electronic products is to process information. As all social relations are predicated on the need to communicate, then electronic products, as I have sugggested on p. 3 have a singularly important utility to almost the entire realm of human activity in contemporary industrial and industrialising societies. Only the production of food, shelter, and clothing can now be seen as more basic to human existence.
3. Electronics production, as we shall discover later (in Chapters 3–6), has generated social and technical divisions of labour, that in combination are quite unlike those of most other manufacturing industries. Specifically, electronics employs relatively large numbers of engineers and technicians and very large (but because of the automation of assembly processes, declining) numbers of unskilled workers. Unlike the steel, automobile, and engineering industries, however, electronics now employs very few skilled manual workers (even taking into account the taylorisation of labour processes in the former industries). Paralleling its peculiar technical division of labour is a particular social division of labour. Specifically, while the overwhelming majority of its scientific, engineering, and technical labour (not to mention senior management) are male, the bulk of its unskilled labour force are young, female, and in many cases migrants (rural-urban migrants in the case of East Asia), or racially or ethnically distinct immigrants (as in the case of the United States).
4. Electronics production, along with a small number of other forms of production (textiles and automobiles for instance), is organised in terms of a combination of 'technically disarticulated' labour processes. Under certain conditions, therefore (which we explore in Chapter 3) particular labour processes can be dispersed to selected locations within the home country or overseas, in order to take advantage of specific combinations of production factors (usually particular forms, qualities, and cost of labour power), or, as in the case of Europe, penetrate the major markets which exist there. With modern telecommunications and transport systems linking dispersed production units to the centralised control function, the 'world factory' phenomenon emerges. More than any other industrial branch, semiconductor manufacture, since the early 1960s, has constituted the prototypical example of a production system organised on the basis of 'world factories'.

These four elements combined, in articulation with historically specific factors such as state-development strategies internal to the given territorial units themselves, have produced the economic, social, and spatial transformations which we analyse in this study. They have produced, for instance, the economic and technological dynamism, but social and spatial polarization evident in the industry's best-known territorial base, Silicon Valley (Saxenian 1981, 1983a, b). They have led, also, to the industrialisation of previously non-industrial parts of core societies, such as in the case of the 'M4 Corridor' in Southern England (Hall 1985; Breheny and McQuaid 1987) and Toulouse and other parts of Southwest France (Pottier 1987). In addition, and especially significant for this study, they may constitute (it is still too early to tell) part of the basis for the reindustrialisation of an older, rapidly deindustrialising society such as Scotland (see Chapter 6), as well as providing a key motor force for the industrialisation of a number of peripheral societies, such as those of East Asia (see Chapters 4 and 5).

Things to come

For the reasons explained above, this study uses the semiconductor industry as its empirical vehicle to expose some of the social and spatial dynamics of industrial change in the contemporary world economy. More particularly, however, it focuses on the American semiconductor industry. It does so for the following reasons:

1. American companies totally dominated world semiconductor production until the early 1980s. Although they had been overtaken by Japan in terms of total semiconductor production by the middle of the decade, as Table 1.1 shows, US companies are still the leading producers of the more technologically advanced products, integrated circuits.
2. American companies were, and still are, the technological powerhouses of the world semiconductor industry. Of the 52 major innovations in semiconductor technology between 1947 and 1981, American companies were responsible for 48 of them, including such key breakthroughs as the transistor, the 'planar' manufacturing process, the integrated circuit and the microprocessor (UNCTC 1986, Annex V: 455).
3. American companies internationalised aspects of their production far earlier than their Japanese or European counterparts, and to this day they dominate production in those territorial units where semiconductor manufacturing takes place, with the exception of Japan, parts of Western Europe (but not Scotland) and the socialist societies (Grunwald and Flamm 1985: Chapter 5; UNCTC 1986). US companies, then, are by far the most

Table 1.1 Semiconductor production: Selected leading national producers, 1985–7 (By value: billions of US $)

National Producer	*1985*	*1986*	(a) *1987*
US			
Discretes	1,460	1,498	1,642
ICs	8,665	9,327	10,428
Total semiconductor	10,396	11,129	12,418
Japan			
Discretes	2,584	2,394	2,444
ICs	8,070	8,429	9,242
Total semiconductor	11,870	12,224	13,227
West Germany			
Discretes	423	412	434
ICs	1,408	1,255	1,435
Total semiconductor	1,945	1,774	1,984
UK			
Discretes	191	191	209
ICs	737	782	895
Total semiconductor	979	1,027	1,159
Italy			
Discretes	129	136	142
ICs	445	417	468
Total semiconductor	599	577	637

Source: Electronics, 22 January 1987.
Note: (a) Estimates.

important semiconductor producers (in terms of both output and employment) in all the territorial units with which this study deals.

The theoretical arena in which this study is set, and to which it hopes to contribute, is that which is concerned with the analysis of the dynamics and consequences of the changing international divison of labour. We begin work in Chapter two, then, with a critical interrogation of some of the principal bodies of literature which seek to explain territorial development, but particularly industrial change as one aspect of that, in both Third World and advanced capitalist societies. Having identified, in the abstract, some of the serious analytic problems embedded in the current literature, we move, in Chapter three, to an account of the development of the US semiconductor industry in its principal domestic base, Santa Clara County, California. There we discuss in concrete form some of the elements of the new mode of industrialisation identified above. We also analyse the determinants of the industry's initial internationalisation of parts of its productive capacity.

In Chapter four we begin our analysis of some of the principal developing societies that have been recipients of US semiconductor investment in labour-intensive assembly processes: those of East Asia. In this chapter, however, we are particularly concerned to analyse the historical evolution of the production processes in that part of the world, in order to explain why some territorial units, rather than others, have been up-graded in terms of their technological bases and control functions, and why, therefore, a distinct regional division of labour seems to have emerged.

In Chapter five we continue the analysis begun in the previous chapter, by examining (in a more rounded fashion than has been possible in our other case studies) the emergence of one of the 'cores' of the East Asian division of labour: Hong Kong.

In Chapter six, we turn our attention to Europe to analyse some of the problems and possibilities associated with the development of its principal semiconductor production complex: the central belt of Scotland. In that chapter we recognise that, historically, Scotland has been a recipient of more technologically advanced labour processes than have any of the East Asian countries and hence it plays a very different role in the industry's international division of labour than any of the latter.

In Chapters four to six, we try to understand the development of semiconductor production in each territorial unit, in relation to both the international division of labour and the internal dynamics of the respective societies. In each case, we try to accord relevant explanatory weight to the respective determinants of development. The methodological basis on which we adjudicate between the determinants is sketched towards the end of Chapter two.

In Chapter seven, we attempt to indicate the principal processes currently affecting the international semiconductor industry in various parts of the globe, including in those territorial units which we have had under particular scrutiny, and indicate the prospects for development that appear to be associated with them. Finally, in the concluding chapter, we reverse the direction of our analytic process in an attempt to assess what our study of semiconductor production and of the territorial units in which it has taken root, have to contribute to the evolving theory of the international division of labour.

Chapter two

The international division of labour, industrial change, and territorial development: Theoretical and methodological issues

In the previous chapter, I argued that one of the key elements which distinguished current attempts to cope with economic crisis was the recourse to spatial shifts in investment, and hence the emergence of depressed deindustrialising regions on the one hand and newly industrialising (or, in the case of Scotland, reindustrialising) areas on the other. The use of international space in this way, for the purposes of new rounds of accumulation, I referred to as the 'global option'. In this chapter, I explore some of the general theoretical issues which underlie attempts to explain the 'global option'. These issues will structure our discussion of the globalisation of semiconductor production developed in Chapters three to six. We will return to them, more abstractly, in the concluding chapter, for my aim in this study is not merely to assess the dynamics of the international semiconductor industry, but also to assess the significance of the empirics of that industry's development, for the theory of the international division of labour.

Territorial development

In seeking to analyse the emergence and evolution of a particular mode of industrialisation in particular parts of the world, it is necessary to understand the spatial form wherein that industrialisation takes root, and which is altered ultimately by its development. This is so because, as Doreen Massey (1984: 51) reminds us, 'the world [does not exist] on the head of a pin . . . it [is not] distanceless and spatially undifferentiated'. Industrial corporations seek to set up production facilities not merely in the United States or Britain or Brazil or East Asia, but in particular spatial locations – in particular cities or regions – within those larger spatial entities. They do so because a particular mix of production factors – types of labour power, finance, infrastructure, local state policies, perhaps – exist there, in a more suitable mix than elsewhere in that same country.

However, it is not simply a matter of a given mix of production factors in a given location, but also of the fact that mix occurs in that location at a particular historical moment. What may be a suitable mix from the point of view of a prospective transnational investor at one moment in time, may have altered significantly ten or twenty years later. Such an alteration may have important consequences for the type of operation (labour processes etc.) the company decides to maintain there, or indeed whether it decides to maintain production facilities at all.

This study deals with industrial change in particular spatial locations. Unfortunately, however, the general traditions of analysis that might help us to understand these processes are, in addition to their other problems, flawed in terms of their concepts of space and time. This is particularly true for the multidisciplinary area that purports to explain economic, social, and political change in particular locations: urban and regional studies.

We begin this chapter, therefore, with a brief identification of the problems of 'traditional' urban and regional analysis. Our dissatisfaction with this work, and particularly its implied concepts of the 'urban' and the 'regional', leads us to propose the concept of the 'territorial unit'. This concept helps us to link spatially specific change with the more general, global processes of change analysed by certain traditions of development theory. The 'territorial unit', then, becomes an heuristic device which helps us to analyse economic and social change in 'cities' and 'regions' in both 'developed' and 'developing' societies in relation to changes in the world economy.

The paradigms which have informed the 'traditional' fragment of urban and regional studies, generally speaking have been rooted in an empiricist epistemology. As a result they have led to accounts of the development of particular cities or regions (or aspects of them) that have been largely descriptive, atheoretical, ahistorical and non-cumulative. Methodologically these accounts have artificially abstracted the city and the region from their larger social and economic contexts and consequently assumed that most, if not all, that was needed to explain urban and regional development was somehow endogenous to the city or region and hence for analytic purposes, they could be treated as closed systems (Forrest, *et al.* 1982). Fortunately, however, the fallacies of this 'ontology of the urban and regional' have now been exposed, and in many arenas overturned. Urban and regional studies have undergone a paradigm shift of revolutionary proportions (Lebas 1982). Thanks to the interventions of Castells (1977, 1978, 1983), Harvey (1973, 1982, 1985), Lefebvre (1976, 1979), Lojkine (1976), Mingione (1981) and many others who have been influenced by them, it is no longer possible to understand the economic, social, political, and spatial aspects of urban and regional development in any intellectually serious way, except by means of a method which emphasizes the embeddedness of the city and the region in their wider structural contexts.

About the same time that urban and regional studies were undergoing this intellectual revolution, development studies were also subject to substantial theoretical shifts. Since the late 1960s a growing body of literature, framed within the 'dependency paradigm', has broken with the ethnocentrism, empiricism and associated theoretical myopia of much traditional social science that claims to analyse 'development' (cf. Baran 1967; Frank 1971, 1980, 1981; Emmanuel 1972; Amin 1974, 1976; Cardoso and Faletto 1979; Wallerstein 1974, 1979, 1980, 1983; Warren 1980). Though there are significant internal differences (cf. Palma 1981), collectively this literature suggests that it is simply not possible to adequately comprehend a vast array of economic, social, and political phenomena in particular Third World countries unless these phenomena are theorised in relation to the changing circumstances under which those countries are structurally connected with the economic and political systems of the advanced capitalist societies.

Though urban and regional studies and development studies seemed to be emerging with similar methodological trajectories, until recently there were very few explicit links between the two fields. The 'new' urban theory tended to be directed primarily to the analysis of cities and regions in advanced capitalist societies (though the work of Castells and Walton were always significant exceptions here), whereas development theory was predominantly concerned with aggregates geographically larger than cities or regions and almost exclusively in Third World societies.[1] The one body of literature, however, that did seem to suggest a possibility of analysing the development of territorial units (including cities and regions) in the 'developed' world in a systematic relationship *both* to each other *and* to equivalent units in the 'developing' world, was that encoded by Wallerstein as 'world-systems analysis'.

The chapter proceeds by examining some of the problems and possibilities that both world-systems analysis, and the theory of the international division of labour associated with it, exhibit for an analysis of territorial development. Specifically, it briefly assesses world-systems analysis in relation to the dependency paradigm out of which it emerged, but by which it remains influenced. It then turns to an extended discussion of the most influential account of the 'new' international division of labour (that by Fröbel, *et al.* 1980). In this section of the chapter I elaborate, with critical comment, on what I take to be the significance of this literature for an explanation of industrialisation and territorial development. In the final section, I comment on the methodological principles that will guide the subsequent analysis.

Dependency theory and world-systems analysis

In order to critically assess the significance of world-systems analysis and the new international division of labour (NIDL) theory for the study of territorial development, it is first necessary to examine some of the problems associated with the dependency paradigm. The reason for this is that although world-systems analysis emerged from the general concerns of dependency theory, in some ways it has continued to exhibit the limitations of the intellectual crucible in which it was forged.

Of the theoretical problems associated with the dependency paradigm in development studies, there are three that are important for our purposes. First, many dependency theories (e.g. the early work of Andre Gunder Frank) suffer from the fact that they locate the problem of Third World 'underdevelopment' within the global circulation of commodities. The perpetuation of underdevelopment, in their view, arises from the fact that capital in its commodity form is transferred from the periphery to the core of the world system on the basis of an unequal exchange (Emmanuel 1972). In more orthodox language, the structural dominance which the core economies possess within the world system (secured in large measure by transnational corporations) results in the terms of trade necessarily being stacked against peripheral economies. These theories, then, put the weight of explanation on the 'circulation' of commodities; on the realisation of surplus value at a global scale (Hoogvelt 1982: Chapter 6).

The result of this theoretical focus is two-fold. First, such theories are unable to explain how unequal exchange is able to be reproduced under changing economic and political conditions; and second, it seems to deny any real possibility for Third World development in the context of a capitalist world economy. On the latter point, it seems to fly in the face of empirical reality by refusing to accord significance to the spectacular growth rates and rising incomes evident in the newly industrialising countries (NICs) in general and the Asian 'gang of four' (South Korea, Taiwan, Hong Kong, and Singapore) in particular (cf. Schiffer 1981; Browett 1985, 1986; Hamilton 1983; Deyo 1987).

For many of the critics of early dependency theory (e.g. Laclau 1977: Chapter 1; Cardoso and Faletto, 1979), the route to a more adequate explanation of uneven development lay in a shift of attention from the problem of the realisation, to the problem of the production of surplus value. What required empirical and theoretical attention, therefore, were the mechanisms by which surplus value was extracted from the periphery by means of the 'super-exploitation' of human labour. This led to a concern with the way in which elements of the capitalist mode of production articulated with economic and social relations emanating from non- (pre-) capitalist modes of production to produce particular class structures and state forms, which, though conducive to continued 'super-exploitation', could theoretically hold out the possibility of harnessing global capitalism in the interests of national development (Laclau 1977, Chapter 1; Foster-Carter 1978; Taylor 1979; Palma 1981).

The second problem of such dependency theory was that as with much 'orthodox' Marxist theory since Lenin, and those brands of neo-Marxism that have been heavily influenced by the work of Althusser, it saw historical change in both core and periphery as largely a product of the working out of the logic of capital accumulation (Aronowitz 1982). More precisely, it appeared to view the 'development of underdevelopment', its changing forms and contents from one historical period to the next, as a product of the vicissitudes of the realisation of surplus value on a global scale. Whatever the state form or social structure of a 'developing' country, for instance, it tended to be viewed as a product of the logic of capital accumulation in one period and functional to continued accumulation in the next. Social classes, races, nation states, seemed to have little autonomy from the global structures of capital, controlled by the core economies. They appeared to have little capacity, therefore, to direct their own destiny, to 'make history'. The only chance for such possibilities arose when the internally generated contradictions of global capitalism became so severe that the whole system was ruptured, thus producing the structural possibility for classes and other social groups to become the agents of historical development.

The third problem with the dependency paradigm, and one to which we have alluded already, was that it was 'only' concerned with development in Third World societies (and even then, largely with macro-level development aggregates). It was not designed to cope with structural changes in core regions or cities in relation to development on the global periphery. However, it is precisely this question that this study, in part (for instance in relation to Scotland), is concerned with.

While Immanuel Wallerstein's version of world-systems theory suffers from many of the same defects of other contributions within the dependency paradigm outlined above (see Brenner 1977; Laclau 1977: 42–50; Skocpol 1977; Worsley 1980; Aronowitz 1981), it does appear to have a number of advantages. First, unique among development theories, it holds out the promise of analysing the internal development of 'socialist' societies in relation to the global development of the capitalist world system (Wallerstein 1979; Chase-Dunn 1982). It is therefore able to make the important point that the possibilities for a socialist route to the 'modern world' in any particular 'developing' society are determined, in part, by the mode and extent of that society's integration in the capitalist world-system both prior and subsequent to its revolutionary rupture.

More significant for our current purposes, however, are two further features of world-systems theory. First, it follows Marx in recognising that the production and realisation of surplus value (circulation of commodities) are but moments of the same accumulation process which must be held methodologically in dialectical relation to each other if any adequate theory of core-periphery relations, uneven development and the rest, is to emerge. World-systems theory, therefore, bridges what Hoogvelt (1982: Chapter 6) refers to as the 'circulationist' and 'productionist' versions of dependency theory, and hence appears to advance the methodological possibility (albeit at a high level of abstraction) of a more 'rounded', comprehensive, and historically based theorisation of global uneven development than perhaps has previously been possible.

In addition, Wallerstein's framework, as with a small number of others (such as, Warren 1980; Schiffer 1981; Marcussen and Torp 1982) recognises the possibility of 'genuine' peripheral development within the capitalist world-system. It does so by the introduction of the concept of 'semiperiphery'. Although Wallerstein's use of this concept remains ambiguous in the sense that the criteria by which it might be possible to designate a country (or city, or region) as semi-peripheral, and the historical processes by which peripheral or core territorial units come to attain semiperipheral status, have yet to be rigorously specified,[2] the concept exhibits analytic promise. First, not only does it express the possibility for capital accumulation in certain parts of the periphery, but it accepts that some of that capital is able to be retained (in the semiperiphery) rather than necessarily being transferred to the core. In spatial terms accumulation in the semiperiphery is multi-directional. Partly by virtue of the emergence of transnational corporations indigenous to certain semiperipheral countries, the semiperiphery is itself able to exploit labour power in the periphery and (as yet to a far lesser extent) the core, and hence 'cream-off' surplus value from both. (cf. Lall 1983). In addition, however, by virtue of foreign investment in production (industrial and/or agricultural depending on the particular case), the semiperiphery as with the periphery, remains a location for the accumulation of capital which is then transferred to the core. Finally, in circumstances of declining profitability in the core (albeit with significant variations from one branch/sector to another) while much capital 'flies' to peripheral regions within the core countries themselves (cf. Massey and Meegan 1982; Bluestone and Harrison 1982, for the British and American experiences respectively) a significant proportion of it in recent years has tended to be 'relocated' to the global semiperiphery and periphery (see, Henderson and Cohen 1982a; Flynn 1984; Gaffikin and Nickson 1985, for details).[3]

From the perspective of this study, at least three problems remain with these essentially spatial concepts of world-system change. First, the core-semiperiphery–periphery configuration are in Wallerstein's work posed merely as formal categories. Although they may designate particular territorial units in particular historical periods, the historical processes, the motor forces that can transform, say, a peripheral unit into a semiperipheral unit, or a semiperipheral unit into a core unit over time are neither specified nor adequately theorised. As Aronowitz (1981) has pointed out, by adopting the concept of historical conjuncture developed by his mentor, Fernand Braudel (cf. Braudel 1980), Wallerstein gives us the impression that economic and social transformations of the sort indicated above, are to be viewed as largely the result of a spatial convergence of a series of historical 'accidents'. As such, Wallerstein's elaboration of the core-semiperiphery–periphery configuration seems strangely discordant with the theoretical tenor of his argument about the development of the world-system as a whole. Second, although the designation of a spatial unit as core, peripheral, or semiperipheral is in part based on the extent to which the unit is the headquarters location for corporations which extract surplus value, or alternatively, of labour forces which have surplus extracted from them, there is in Wallerstein's conceptualisation an inadequate understanding of the role of organisational and scientific/technical (knowledge) control in the core-semiperipheral–peripheral configuration. The assumption seems to be that core economies are the locus of all control of these varieties. The possibility of semiperipheral units operating as 'way-stations' in the control structure of the world economy, and as such having a capacity themselves to exercise a modicum of organisational and technical control, seems to have escaped Wallerstein's attention. As we shall see in our discussion of semiconductor production in East Asia (Chapters 4 and 5), this point takes on particular significance for some of the territorial units in that region.

The final problem with Wallerstein's work, from the point of view of this study, is that it pays insufficient attention to a central element in the dynamics of the world-system, namely the nature, distribution, and interrelationship of labour processes across the globe. That problem, however, is addressed more directly by another body of literature associated with the world-systems tradition, which, following Fröbel *et al.* (1980), has come to be known as the new international division of labour (NIDL) thesis.

The 'new' international division of labour

What is new about the 'new' international division of labour?

From its beginnings in early sixteenth-century Europe, the capitalist world can be conceived as having developed through three phases of the international division of labour (Walton 1985b). In its first phase that division of labour was characterised by the extraction of minerals and agricultural produce from the periphery by means of the forcible application of human labour power (in forms such as slavery) to production. In addition to being the locus of mercantile and military control for the entire system, the core units themselves engaged in agricultural, mineral, and small-commodity production, and traded largely with other core units. The second phase, whose high point coincided with 'classical' imperialism of the nineteenth and first half of the twentieth centuries, involved the development of industrial production in the core, and the trading of industrial and agricultural commodities between core and periphery, with the periphery continuing to concentrate (supposedly by virtue of its 'comparative advantage') on agriculture and minerals. In this phase, the extraction of surplus from the periphery was largely via the sphere of circulation. As indicated earlier, it was predominantly the sphere of circulation that has been the theoretical focus of most Marxist and neo-Marxist accounts of uneven development.

Beginning with the work of Christian Palloix (1977, first published in French in 1975), however, it began to be recognised that the international division of labour, and with it the mechanisms of capital accumulation, had entered a new phase. It appeared that the 'self-expansion' of industrial capital could no longer be achieved solely within the national boundaries of the core economies of the USA and Western Europe themselves. The internationalisation of productive capital had been evident, of course, for some considerable period, and manufacturing capital from one core economy had been invested in production facilities in other core economies at least since the turn of the century (automobiles being, perhaps, the classic example). What has been 'new' in the last quarter century, however, has been (a) a vast increase in foreign direct manufacturing investment in the core economies themselves (Läpple 1985) and (b) foreign direct manufacturing investment, usually for the first time, and on a significant scale in certain 'Third World' or 'peripheral' societies. The latter process has helped to create what are currently among the world's most productive economies, the so-called 'newly industrialising countries' (Browett 1986; Deyo 1987). Concomitant with these developments, manufacturing activity in many parts of the world has come to be increasingly organised and co-ordinated at a global level, largely by transnational corporations (Hymer 1979; Palloix 1977).

While capital was being exported to the periphery in search of the application of Third World labour power to particular labour processes, so the 'free' migration of peripheral labour power (a motor of the accumulation of industrial capital from its eighteenth-century beginnings) continued to be applied to particular labour processes in the core, in spite of juridico-political restrictions on periphery-core migration from about the early 1960s (Cohen 1987a). Because the global mechanisms of capital accumulation had entered this new phase, Palloix argued that the analytic focus had to be shifted from the sphere of circulation to the sphere of production, from the problem of the global realisation of surplus value to the problem of the globalisation of its production (Palloix 1977).

NIDL theory: its strengths and limitations

Though the object of increasingly critical attention (see for instance, Jacobson *et al.* 1979; Jenkins 1984; Cohen 1987b) as with others who have studied the internationalisation of industrial branches (e.g. Hill, 1987; Sayer, 1986a), I utilise the most developed account of the empirics and significance of the NIDL (as far as manufacturing activity is concerned[4]) so far available – by Fröbel, *et al.* (1980) – as an heuristic device from which my argument will emerge. The essence of the account developed by Fröbel, *et al.* (1980) can be summarised as follows:

1. The migration of industrial capital from the core to the periphery of the world-system has increased significantly in recent years, and has resulted in the development in a number of Third World countries of industrial enclaves composed of 'world-market factories', which manufacture commodities (often partially finished) primarily for export.[5] This migration of capital has been spurred on by a gradual worsening of the conditions for valorisation in the advanced capitalist societies (partly because of successful industrial conflict; see, Bluestone and Harrison 1982), together with the existence at the periphery of seemingly endless

supplies of low-wage, unorganised labour. There now exists an industrial reserve army of global proportions, which in recent years, given increasing national restrictions on the import of labour in many core economies (e.g. EEC), can best be exploited by the export of capital.

2. According to the NIDL thesis, the emergence of this global reserve army is relatively recent and has resulted from the increasing penetration of commodity relations into Third World agriculture. As a consequence, traditional social and economic structures have been undermined and proletarianisation, under-employment, and unemployment have advanced on a massive scale. This in its turn has led to an urban migration of the rural poor (Roberts 1978; Portes and Walton 1981). In addition, the retention of forms of social reproduction and non-wage remuneration in households (informal economies) has helped to reduce pressure for rising industrial wages that might otherwise have been expected (Wallerstein 1978, 1983: 69).[6]
3. Developments in the labour process, particularly those associated with Taylorism and Fordism have resulted in an increasingly minute technical division of labour which has allowed an organisational and ultimately spatial separation of particular labour processes within the same industrial branch and firm. Generally, the more deskilled and labour-intensive processes have been relocated to the periphery to take advantage of cheap labour, whilst the more skilled and technologically advanced processes have been retained in the core economies. This relocation has resulted in a reduction of manual employment opportunities in the core, which in the context of deepening economic crisis in those economies, has contributed to rising unemployment (Massey and Meegan 1982). The development since Taylor's day of increasingly refined forms of 'scientific management' has ensured that labour productivity in world-market factories now equals or exceeds that in many core economies.
4. There has been a provision of incentives by international agencies (e.g. UNIDO, World Bank) and national governments to attract industrial investment to the periphery. The latter, in particular, have been involved in the construction of export-processing zones, the institution of 'favourable' labour laws involving, for instance, either the outright suppression, or at least considerable restrictions, on trade union activity, tax holidays and credits for foreign corporate investors, freedom to repatriate profits, infrastructure provision, negligible industrial health and safety regulations, freedom from planning and environmental controls, etc.

5. The consequences of all the above at the level of the world economy is that new possibilities for the extraction of absolute surplus value have been created by virtue of the 'super-exploitation' of Third World labour. As a result, the globalisation of industrial production in this manner has helped to moderate one of the determinants of economic crisis, the tendency of the rate of profits to fall.[7]
6. The spatial dispersal of particular labour processes within the same industrial branch or firm has been facilitated by innovations in transport and communications technologies. The introduction of containerised shipping and efficient air cargo systems for instance, has been of major significance, as has been the development of electronic information systems.
7. The analysis of Fröbel *et al.* (1980) leads to very pessimistic conclusions regarding the possibilities of genuine development in the periphery arising from export-oriented industrialisation strategies.[8] They suggest, for instance, that these strategies result in the transfer of technologies that are either outmoded in core economies, or, at best, other than 'state-of-the-art'. In addition, because the technologies transferred have deskilled labour processes associated with them, they do little to improve the skills of peripheral labour forces. As world-market factories tend to employ a predominance of young women, their presence does little to alleviate male un- and under-employment. The consequence of these two elements is that they help to generate badly skewed labour market structures constituted by male unemployment, a large component of unskilled female employment (particularly when the service and domestic sectors also are taken into account) and a smaller (but not insubstantial) component of white-collar and professional labour. Furthermore, they suggest by reference to those countries that have been incorporated into the NIDL the longest (e.g. Hong Kong and Taiwan), that this situation does not appear to improve over time. Finally, they argue that world-market factories have few links with local economies. Most of their purchases involve intra-firm transactions or purchases from other foreign-owned corporations, and their only benefit to the local economies are their limited purchases of simple supplies (e.g. packaging for shipment) and utilities (electricity, gas) and their stimulation of local commodity sales by virtue of the wages they pay their workers.
8. In sum, though the forms of industrialisation grasped by the NIDL thesis may constitute development *in* a country, city or region, it has little chance of assisting development *of* that territorial unit.

Though many particular aspects of the analysis of Fröbel and his colleagues have been confirmed by earlier and subsequent case studies of particular industrial branches and/or territorial units (e.g. Evans 1979; Evans and Timberlake 1980; Lim 1978a, 1978b; Sunoo 1978, 1983; Portes and Walton 1981; Luther 1978; Ip 1983), there are a number of problems that are of particular interest to this study:

(i) At the most general level the analysis of Fröbel and his colleagues suffers, like much neo-Marxist development theory, from the fact that it posits a capital-logic approach to uneven development. In their case this approach is insistently argued:

> The new international division of labour is an 'institutional' innovation of capital itself, necessitated by changed conditions, and not the result of changed development strategies by individual countries or options freely decided upon by so-called multinational companies.
>
> (Fröbel, *et al.* 1980: 46)

As a result, their work tends to replicate the theoretical error of devaluing the role of nation states, classes, and other social forces (racial or ethnic groups, for instance) in the development trajectories of particular countries. It is not that their account is wrong necessarily, but rather that it is heavily one-sided and simplistic. By failing to accord any degree of autonomy to nation states (be they in the core or periphery), classes/class fractions (be they bourgeois, proletarian, or 'semi-proletarian') or other social groups, they appear to rule out the possibilities of 'national' capitalist development within the world-system, in spite of what now must be taken as significant empirical evidence to the contrary.

(ii) A second and related problem is that in spite of the wealth of empirical information which they bring to bear on the analysis, their account of the contours and significance of the NIDL still appears to be pitched at too abstract a level. While it is the case that the possibilities for development in a particular territorial unit are structurally determined by the mode and extent of that unit's integration in the world-system, the root of the problem, I suspect, is two-fold. First it lies in their implied notion of 'determination'. They appear to operate with a mechanistic, and indeed discredited concept of the 'superstructure-as-a-reflection-of-the-base' variety, albeit in this case, spun-out at the global level. A more satisfactory concept of determination is one that sees economic relations as 'determining' political processes and class formation, for instance, in the sense of setting structural limits to, and imposing pressure on, their possibilities for variation (cf. Wright 1978: Chapter 1; Williams 1980). Applied at the global level, such a concept of determination while probably logically ruling out the possibility of a semiperipheral unit achieving core status (at least in the context of the contemporary world system), would allow for the material and social transformations which have constituted (and for their populations *feel* as though they have constituted) genuine development in such semiperipheral units as Hong Kong and Singapore for instance.

(iii) Following from the previous point, it can be argued (as, for instance, Palma (1981) does for dependency theory) that Fröbel *et al.* (1980) are insufficiently sensitive to the empirical specificities of 'development'. Especially important here is their tendency to under-value the impact and development of production facilities by particular industrial branches or firms in particular territorial units. Their almost exclusive concentration on the supposed relationship between the internationalisation of industrial capital and the search for new opportunities for the extraction of absolute surplus value, for instance, causes them to miss the significance of changes in the labour process, in capital-labour ratios etc. which may occur in particular peripheral locations *subsequent* to the initial investment of a particular firm or branch in productive facilities. Consequently they fail to grasp the point that as a result of accumulation imperatives internal to the international organisation of the firm itself, or pressure from national states, a concern to increase relative surplus value in particular production locations may develop over time.[9] A concern with relative surplus value, of course, can result in rising real wages, 'advanced' technological transfers etc. In addition, it can lead to a greater integration between foreign-owned production facilities and local economies. In the best of circumstances, those peripheral operations of foreign firms that are organised around the question of relative surplus value can help to stimulate the development of local production complexes. The case studies presented here (Chapters 4–6) partly explore precisely this phenomenon.

(iv) The criticism contained in the previous paragraph can be developed further. Fröbel and his colleagues seem to assume that whatever the origins of the internationalisation of production within one branch or firm, the course of the subsequent development of the internationalisation strategy remains much the same, irrespective of the particular branch or firm one is concerned with. The corollary of this is that they assume that the economic, technical and social impact of a given branch or firm tends to be much the same, irrespective of the particular peripheral country in which world-market factories are established. The account of the NIDL which Fröbel *et al.* provide, therefore, is both profoundly ahistorical and insensitive to the differences in the internal balance of social forces in particular national states, and hence to differences in their development strategies. As a result, Fröbel and his colleagues fail to recognise that production strategies of firms change over time subsequent to the initial internationalisation, and that they articulate in different ways, in different periods, with the economic, political, and social circumstances of particular 'developing' or 'developed' territorial units. Put another way, while the form and content of the NIDL may be much the same from one branch or firm to another at the moment of initial internationalisation, both of these may well change over time, depending on the branch or firm one is concerned with. If this is indeed the case, then it may pose significant implications for the role of particular territorial units within the NIDL.

(v) Another aspect of this whole question is that by failing to focus on the history of productive operations of particular firms and industrial branches in particular locations, Fröbel *et al.* miss the fact that rather than there being one international division of labour, a number of related spatial divisions of labour may be emerging within certain industrial branches. In Chapter four, I argue that this phenomenon, in fact, is occurring within US semiconductor production in the East Asian region, where places such as Hong Kong and Singapore have emerged as the 'cores' of a regional division of labour, with Malaysia, Thailand, the Philippines and Indonesia subordinate to them. Such phenomena as these again potentially have enormous significance for the development possibilities in each case.

(vi) Finally, we need to comment on the type of information that is required to assess the relative prospects for development within the NIDL. This information is not only that which details changes in GDP, or real wages, or welfare expenditure etc. What is also required are 'theorised histories' (Castells 1977: 437–71) of particular industrial branches (perhaps particular firms) operating in particular territorial units, and their relation to class and state formation (with their own experiences of struggle, development priorities) etc. In addition these theorised histories must be comparatively developed, as for instance Castells (1983) has done in his recent work on urban social conflict and Armstrong and McGee (1985) have done in their work on Asian and Latin American urbanisation. It seems to me that it is only in this way that we can discover the actual labour processes, technological inputs, connections with local economies, impacts on labour market structure, class formation, and the rest that are the consequence of particular forms of capital investment in both 'developed' and 'developing' societies. Only when we have such theorised histories at hand can we begin to adequately assess the possibilities for development, be it in a city, country, or region, within the capitalist world-system.

As this study progresses, a number of these problems with NIDL theory will be picked up and elaborated with regard to semiconductor manufacture in particular territorial units. Before turning to discuss some methodological issues of concern to our study, however, we need to comment, in passing, on another recent attempt to explain the spatial organisation of industrial production.

Building on the earlier work of Vernon (1966), Ann Markusen (1985) has proposed a 'product-profit cycle' model to explain the territorial arrangement of manufacturing activity. She suggests that:

> regional shifts in production and employment (are substantially the product of) disparate strategies undertaken by corporations experiencing different historical moments of longer-term profitability cycles. Regional shifts in the aggregate are composed of two major impulses, the tendency to rationalize and cheapen production on the part of more mature, profit-squeezed sectors and the tendency to innovate and concentrate production at virgin sites by young, superprofit sectors. The resulting regional adjustment problems are compounded by oligopolistic practices.
> (Markusen 1985: 1–2)

While the profit-product cycle model might help us to overcome one of the difficulties posed by Fröbel *et al.*'s NIDL thesis – namely their ahistorical account of the development of production facilities in particular territorial units, from our point of view, it still has at least two major weaknesses:

(i) Its explanation of the spatial distribution of production, and changes thereof, is strongly mono-causal in the sense that the state of the product's technological evolution (mature or innovatory) together with the questions of profitability in part associated with it, is seen as determining the spatial organisation of the firm or branch. Markusen's account is, then, highly mechanistic and fails to comprehend the significance of social and political factors indigenous to the territorial unit and their historical development. In this sense, therefore, her account is in some ways as one-sided as that developed by Fröbel and his colleagues.

(ii) Unlike the NIDL thesis, the product-profit cycle model fails to distinguish between the technologies, capital–labour ratios, labour power requirements associated with particular labour processes. It deals only at the level of whole production systems. As a result, it cannot explain the spatial organisation of industrial branches which disperse only particular (usually intermediate) labour processes, with varied sets of labour requirements, profit calculations etc. associated with them. These difficulties become especially clear when the model is applied to semiconductor production (as in Markusen 1985: 109–17) and high technology industry generally (as in Markusen *et al.* 1986). In short, the product-profit cycle model would appear to have major problems accounting for the world factory phenomenon, and therefore, is of little help to us in this study.

Before we can begin our substantive work on the globalisation of semiconductor production, I need to indicate, in what remains of this chapter, the methodological principles which will guide the subsequent analysis.

Identifying the determinants of globalisation: A note on method

If a reconstructed NIDL thesis provides a useful starting-point for the explanation of the globalisation of semiconductor production, what in particular are the elements that might be seen to determine the course of that development?

Understanding the meaning and logic of the development of an international division of labour within any industrial branch necessitates attention to (a) the structure of capitalist commodity production at large, (b) the central dynamics of labour processes in the context of the contradictory capital–labour relation, (c) the internal and external organisation of industrial production, (d) the influence of state development strategies, (e) the socio-spatial consequences of these phenomena, and (f) the way this ensemble of relations changes over time. In its attempt to explain the globalisation and developmental significance of American semiconductor production, this study works with these determinants in their various combinations, depending on geographic location and the historical phase of their development. Chapters three to six are attempts to give empirical 'life', as it were, to these determinants in time and space. But before we begin, we need to have some sense of the relative primacies to accord the various determinants in our explanatory system. Without some means of distinguishing their respective explanatory 'weights', we would be in danger of elaborating a theoretically incoherent account of the spatial development of semiconductor production.

In order to avoid such an outcome, we need to take seriously Sayer's (1982, 1984) distinction between relations that are internal and necessary to the development of a phenomenon, and those that are external and contingent. This does not mean, of course, that externally contingent relations are to be treated as if they were of no account in social scientific explanation. On the contrary, such relations are of importance in the analysis of particular empirical outcomes. What it does mean, however, is that these relations do not constitute the primary determinants of the phenomenon in general. Rather, their significance arises when they articulate with the primary determinants (the internally necessary relations) in particular empirical (and historical) contexts.

For an analysis of the spatial development and social impact of semiconductor production (or any industrial branch), this methodological prescription implies that its function as a part of capitalist commodity production at large is the essential starting-point, and hence relations associated with valorisation (production of value) and commodity circulation (realisation of value) are crucial.

In connection with the problem of commodity circulation, the structure of the market becomes an important determinant of both the spatial dispersal of an industry and its subsequent evolution in any particular location. With regard to the problems of valorisation, the nature of the labour process and of associated phenomena such as capital–labour ratios, technological change, wage rates, and actual or potential labour conflict, are central issues. As Sayer (1984: 82–4) recognises, however, internally related objects do not always have to be equal parties to the relation. They can be asymmetrically related so that one object can exist without the other, but not vice versa. Thus, to take an example pertinent to this study, while capital accumulation can exist without state intervention, the state cannot exist without capital accumulation (as this is the basis of its revenues). Additionally, contingently related objects can become central to the existence of a phenomenon when that phenomenon assumes a particular concrete form. Thus, while state policy may not have been central to accumulation during the epoch of competitive capitalism, its transformation to the monopoly or 'late' capitalist form seems to have been intimately associated with state intervention (cf. Habermas 1973; Offe 1975).

Throughout this study, then, while markets and labour processes etc. will be treated as relations internally necessary to semiconductor production, the role of the state, in its various guises, will be seen also as a significant factor, helping to determine both the globalisation of production and its impact in particular territorial units. In the light of the logic sketched above, however, methodologically the state must be viewed as a contingent factor with regard to the general, transhistorical phenomenon of economic change. When it comes to economic change in some of its particular, historically specific forms, however, state policy becomes a centrally important (perhaps the most important) contingency. The globalisation of high technology production and its economic and social impact is a case in point.

Chapter three

Semiconductor production: Labour processes, markets, and the determinants of globalisation

In this chapter, I lay the foundations for our subsequent examination of semiconductor production in East Asia, Hong Kong, and Scotland (Chapters 4, 5, and 6 respectively). These foundations have a number of elements. First, I briefly explain the nature of the semiconductor industry and its product in as much technical detail as the reader will require in order to fully comprehend the subsequent argument. Second, drawing on the methodological orientation outlined towards the end of the previous chapter, I examine the sets of relations internally necessary to semiconductor production. Given, as I have already argued, that these internal relations stem from the fact that semiconductor production is part of capitalist commodity production at large, we focus on the problems of the production and realisation of surplus value, and hence on the industry's labour processes and markets. These we take to be the primary determinants of the industry's territorial organisation. However, consistent with our argument at the end of the previous chapter, it will be clear that the activities of particular national states (in terms of military purchases, tariff policies etc.) also have had important implications for the global evolution of semiconductor production.

These determinants must be given historical and empirical 'life', and this we do by showing how they articulated with other contingent factors to produce the world's first major semiconductor production complex in California's Santa Clara County. Having assessed some of the social and spatial consequences of the emergence and development of the industry in Santa Clara County, we extend our discussion to the processes which impelled the industry's internationalisation from the early 1960s. We end the chapter by describing the shape of the international division of labour which US semiconductor companies had produced by the mid- to late 1970s. By so doing, we set the scene for our subsequent discussion of particular territorial units within that division of labour.

The industry and its product

The precursor of modern semiconductors was the vacuum tube. This device, invented in England shortly after the turn of the century, consisted of from two to five or more electrodes surrounded by a glass tube. The tube was sealed to a metallic base and a vacuum was induced inside it so as to improve conductivity. Following the development of wireless telegraphy, the vacuum tube (or 'valve'), became central to the development of all electronics for much of the following half century.

Though it could be easily and cheaply manufactured, the vacuum tube was fragile, bulky, relatively slow in transmitting electronic signals, and because of its incandescent filaments, consumed relatively large amounts of electricity. By the middle of the 1950s, partly because of these disadvantages, the vacuum tube began to be replaced by the semiconductor, the device on which the subsequent 'microelectronics revolution' was to be based. The semiconductor emerged technologically from developments in solid-state physics during the 1920s and 1930s, and its first generation was constituted by the transistor.

The transistor had been invented in the United States shortly after the Second World War. It was first made in commercial quantities by etching circuits on germanium (and subsequently silicon) wafers by means of electrolyte jets. Individual circuits were then cut from the wafer, and manually (using microscopes and soldering irons) attached to wires (which facilitated their connection to other transistors), and then sealed (bounded) in a casing of non-conducting material.

The first generation transistors were relatively unreliable, could cope with only low frequency signals, and the fabricating process produced very low yields. By the late 1950s, a process of wafer fabrication, the 'planar' process, had been developed which was subsequently to help revolutionize, once more, semiconductor technology. The planar process used photolithography to transfer the circuits traced on celluloid filaments ('masks'), and etch them on to the surface of the silicon wafer. The wafer was then subjected to a chemical diffusion process which transferred certain impurities possessing particular electronic qualities to the surface of the wafer. Though the planar process helped to increase yields (i.e. successfully fabricated wafers and circuits) and improve reliability, the assembly process (i.e. wire attachment and bonding) remained unaltered.

The revolutionary character of the planar process arose from the fact that it technologically allowed not only for the possibility of producing many separate circuits from the same silicon wafer, but also for creating many circuits on the same wafer fragment. As a result it constituted an essential precondition for the development of those semiconductors that are the principal building blocks of electronic machines to this day: integrated circuits.

As its name suggests, the integrated circuit ('silicon chip') consists of a multiplicity of circuits, etched on a single fragment of the silicon wafer. Integrated circuits have relatively low power consumption, are highly reliable, cheap to produce, and most importantly are tiny devices which perform more quickly the functions of earlier generations of electronic machines.

Since their first commercial development in the early 1960s, integrated circuit technology has become increasingly sophisticated in two senses. First, the number of circuits that can be carried by a single chip and hence the number of functions that the integrated circuit can perform have expanded enormously. With first 'large scale' (LSI) and then 'very large scale' integration (VLSI) in the 1970s, each integrated circuit now contains in excess of 100,000 microscopic transistors. Second, the range of functions that the integrated circuit can perform has advanced significantly. There are now five basic forms of semiconductor technology. These are, from the simplest to the most complex:

- linear technology – embodied in transistors and other 'discrete' devices, and used, for instance, in telephone systems
- logic/digital technology – used in micro-computers, VCRs.
- memory systems – used in computers, VCRs.
- optical systems – used in TVs.
- microprocessors – the technological core of computers, aerospace and satellite technology.

In addition there are variations in the sophistication of the ranges of technology embodied within each of these basic forms. Thus, for instance, contemporary memory technology ranges from the simplest 16K to the latest generation, 1-megabyte integrated circuits (i.e. chips capable of storing one million pieces of information each). Similarly, microprocessor technology has developed from its initial phase (in 1971) when 4-bit microprocessors were produced to the current 32-bit products. The significance of these different levels of technology embraced by the international semiconductor industry, is that they have varying requirements in terms of production techniques, quality of scientific and engineering labour, etc. Consequently, not all national semiconductor industries or companies are capable of producing the most sophisticated forms. Indeed, to this day, American firms dominate the commercial production of 32-bit microprocessors, and recent reports (e.g. Ernst 1987), suggest that Japanese manufacturers, for instance, will not develop a 32-bit capacity (at least without American technological assistance) until the 1990s. Such technological issues, then, clearly have important consequences for the development of semiconductor production in particular territorial units. They are issues that we will need to return to throughout the study.

Although the continued development of integrated circuit technology has had enormous implications for research and development work and for the masking and diffusion of silicon wafers, until recently it had few consequences for the assembly and final testing of semiconductors. Only in the last few years, partly in response to the evolution of VLSI technology, have assembly processes begun to be automated, and final testing begun to be transferred to the control of lasers and computers, (cf. Borrus, *et al.* 1982; Wilson, *et al.* 1980). As we shall see in the following chapters, these advances in production (as distinct from product) technology have had important consequences also for the industry's differential evolution and impact in various locations.

At this point we move on to examine the internal dynamics of the semiconductor industry. We begin with a discussion of its central necessary internal relations, (that is, labour processes and technical change[1]) and proceed with an examination of its evolution in relation to the various socio-territorial circumstances that have attented its overall development.

Labour processes and technical change

The earliest growth of the solid-state semiconductor industry occurred in the years immediately following the Second World War. As the industry emerged technologically from vacuum tube production, the initial site of this growth was the Northeast of the United States. It was there that the transistor was invented at AT & T's Bell Laboratories in 1947, and first commercially manufactured by another AT & T subsidiary, Western Electric in 1951 (Braun and Macdonald 1982). Throughout the 1950s more than half of all semiconductor production in the United States was concentrated in the New England and Middle Atlantic states (see Table 3.1). A variety of factors, however, combined to discourage the continued concentration of the industry in the Northeast and to stimulate its rapid development in Santa Clara County, California (with secondary centres in Texas and Arizona). I shall discuss these factors further on p. 35–6, but for the present, I am concerned to examine technical change in the industry in relation to labour processes and the structure of the labour force.

Table 3.1 Regional Origin of US semiconductor shipments 1958–72 (percentages)

Region	*1958*	*1964*	*1967*	*1972*
New England	23	17		17
Middle Atlantic	36	31	43	21
New Jersey	10	4		7
New York	13	10		1
Pennsylvania	13	17		13
North Central				3
South	28	25	22	23
West	13	27	35	37
California				23
Other				14
US totals	100	100	100	100

Source: United States Department of Commerce (1979: 7).

The growth of the semiconductor industry in Santa Clara County from the late 1950s coincided with the almost total displacement of the vacuum tube by transistors and diodes in both military and commercial markets. This technological shift not only had enormous implications for the size, cost, and reliability of various forms of electronic hardware, but also for the type of labour processes associated with production. Semiconductor labour processes, though (of course) functionally integrated, were (and are) *technically disarticulated* (that is, separated from one another as relatively discrete clusters of work tasks). Though, under particular conditions, they may be vertically integrated in the same location, under different conditions, the reverse may be true. They are, in other words, potentially spatially separable.

Semiconductor production involves five component labour processes (Figure 3.1). These are (a) research and development (R & D), (b) mask making (production of the celluloid filaments that contain the microscopic electronic circuits), (c) wafer fabrication (process by which the circuits on the mask are transferred to the silicon wafer and etched into its surface), (d) assembly of transistors, diodes and integrated circuits and (e) final testing of the product. These

Figure 3.1 Semiconductor labour processes

Labour process	*Capital intensity*	*Skill requirements*		
		Scientific/ engineering	*Technical*	*Semi-/Unskilled manual*
Research and development	+	+	–	–
Mask making	+	+	+	–
Wafer fabrication	+	+	+	+/–
Assembly	+/–[(1)]	–	+/–	+
Final testing	+	+/–	+	+/–

Note: (1) Where automation has taken place.

component labour processes have widely varying needs in terms of capital investment, labour skills, specialized inputs, and so on. R & D, though capital intensive, also demands the application of highly qualified and creative scientists and engineers. Mask making and wafer fabrication are capital intensive also, and require highly skilled engineering and technical labour. However, the latter is distinguished from the former by the fact that it requires, in addition, significant numbers of semi-skilled workers, particularly as operators of the 'diffusion' furnaces. Assembly has been predominantly a labour-intensive process given over almost entirely to unskilled labour, although in recent years automation has begun to take place. Final testing, though increasingly capital intensive and requiring significant inputs of technical labour, also requires reasonable amounts of unskilled labour.

From the beginning, then, the semiconductor industry generated a marked polarisation in the skill structure of its labour force. This polarisation was compounded by subsequent technological innovations which revolutionized the whole organisational structure of the industry during the 1960s. Of particular importance have been the introduction of (a) the planar process for wafer fabrication (first developed by Fairchild Semiconductor in 1958). Because it facilitates the superimposition of one circuit over another on the wafer, this process was a necessary precondition for the successful commercial development of (b) the integrated circuit (first produced in quantity by Fairchild and Texas Instruments in 1962–3). By the end of the 1960s, the integrated circuit had substantially replaced the transistor as the basic building block of electronic machines. It accordingly became the staple output on which the spectacular success of the semiconductor industry was based. Subsequent developments (in the

1970s) such as (c) the production of microprocessors (first developed by Intel in 1971) and (d) the move to large-scale and then very large-scale integrated circuitry have had additional implications for the ratios of technical to unskilled labour power in the industry. The net result of these developments, has been an intensification of labour market segmentation in the industry as a whole. Hence, the proportion of professional/technical workers has tended to increase over time, while the proportion of semiskilled and unskilled workers has tended to remain fairly stable. By contrast, the phenomenon of a 'disappearing middle' of skilled production workers is increasingly evident. Some of these trends are highlighted in Table 3.2 which shows the occupational structure of the electronic components and equipment industry in Santa Clara County (of which semiconductors form a significant part) over the 1950s and 1960s.

Table 3.2 Occupational structure of the electronic components and equipment industry in Santa Clara County, California 1950–70

Occupational group and gender of employees	*Percentage of total employment*		
	1950	*1960*	*1970*
Non-production workers:	38.5	55.8	59.2
Professional/Technical	19.4	34.5	34.7
women	0.9	2.8	4.0
men	18.5	31.7	30.7
Managerial/Sales	4.5	7.0	8.7
women	–	0.5	0.5
men	4.5	6.5	8.2
Clerical	14.6	14.3	15.8
women	8.7	9.8	11.9
Production workers	61.5	44.2	40.8
Skilled/Supervisors	27.1	17.8	11.5
women	0.7	1.0	1.1
men	26.4	16.8	10.4
Semi-skilled/Unskilled	34.3	26.4	29.3
women	8.3	14.9	19.2
men	26.0	11.5	10.1
Totals	100.0	100.0	100.0

Source: Keller (1981: 95).

This 'peculiar' technical division of labour within the industry has led, over the last couple of decades, to the emergence of socially and spatially segregated labour markets. These segregated labour markets are also marked by a particular social division of labour in terms of gender, race, and ethnicity. Specifically, white males have tended to fill vacancies for scientists, engineers, and technicians, while non-white females have predominated in the market for unskilled production labour. At the same time, the structure of the industry's labour market has had implications for the patterns of urban development and for the contradictions associated with it. These processes have had their most obvious impact in the industry's core complex, Santa Clara County, but to a greater or lesser extent have been played out across the globe subsequent to the development of offshore production.

I have indicated some of the features of the labour processes associated with semiconductor production. I need to add that a number of scholars, building on Marx's insights on the significance of the gradual replacement of the formal by the real subsumption of labour by capital (see, for instance, Aronowitz 1978), have argued that capitalist industrial development necessarily involves regular attempts to reorganise and streamline the labour process. These attempts seek both to lower wage costs and increase productivity, and at the same time to eliminate, or at least neutralise, oppositional elements in the workforce and strengthen managerial control over production (see, amongst others, Tronti 1972; Braverman 1974; Friedman 1977; Edwards 1979; Henderson and Cohen 1979; Nichols 1980; Thompson 1983).

Attempts to reconstruct labour processes have typically involved organisational changes such as Taylorism and Fordism and/or the application of new technologies. Although both these forms of managerial intervention have been present in the semiconductor industry, as Keller (1981) has shown, the latter (for instance in the form of computer-controlled automatic bonding machines) has tended thus far to dominate. This is scarcely surprising in an industry as dependent for its success on technological innovation as the semiconductor industry has been. These developments have also had important implications for the spatial development of the industry. It is to these questions, initially in terms of the emergence of the industry's primary location – 'Silicon Valley' – that we now turn.

The formation of the core complex: The silicon valley phenomenon

It is often the case that the spatial development of an industry is broadly divisible into two major phases, namely, a first phase characterised by spatial convergence and concentration of producers, and a second phase characterised by decentralisation (Norton and

Rees 1979; Scott 1980, 1982). The American semiconductor industry is no exception to this general tendency. In the present section, we deal with the dynamics and manifestations of the first phase of development in the semiconductor industry. This involves, above all, an analysis of the steady build-up of semiconductor plants and ancillary services in the Santa Clara Valley of California from the mid-1950s to the early 1970s. In subsequent sections of this chapter, and in later chapters, we consider the industry's second phase which has been marked by a massive decentralization of production facilities on both a national and a world scale.

The emergence of the semiconductor industry in California

In the previous section, I mentioned that the growth of the semiconductor industry coincided with a locational shift from the northeastern states to Santa Clara County. There were a number of reasons for this shift. In the first place, most of the large electrical and electronic apparatus producers in the Northeast were already by the early 1950s strongly unionised (Troutman 1980; Bluestone and Harrison 1982). There was little to prevent the eventual incursion of union organisation into the semiconductor sector had it finally settled in the Northeast.[2] Second, as a result of anti-trust law suits brought by the Federal Government, AT & T, the original producers of the semiconductor, were forced to liberally license their technology. In this way, then, small firms, such as Texas Instruments (in Dallas) were able to develop their own productive capacity, as were the Japanese, for AT & T did not restrict its licensing agreements to domestic producers (Grunwald and Flamm 1985: Chapter 3). Third, by the mid-1950s the entire US aerospace-defence industry was growing with great rapidity in the Southwest of the country, and it started to draw into its geographical orbit a surrounding constellation of high technology industries, including the semiconductor industry.

Thus, when William Shockley (one of the inventors of the transistor) established the first semiconductor plant in Santa Clara County, just south of San Francisco, in 1955, he alighted on a location where the prospects for the new industry were especially bright. Among other advantages of the location were the nearby aircraft and other militarily related industries, and the large teaching and research centres of Stanford University and the University of California, Berkeley, which provided trained scientists and engineers, essential to the more technologically advanced sectors of semiconductor production. A note of methodological caution is necessary at once, however. Much of the literature on the development of the semiconductor industry in Santa Clara County puts great explanatory emphasis on Shockley's locational decision and on the proximity of major universities (cf. Bernstein, *et al.* 1977; Braun and Macdonald 1982; Saxenian 1983a, 1983b). We must be careful in ascribing causal powers to these phenomena. Shockley certainly planted the seed of the future development of what came to be known as Silicon Valley, but there can be no guarantee that any initial decision of this sort will come to social and economic fruition. Such fruition depends upon a series of additional necessary conditions of development. Although it was the case that Stanford University had developed what would today be called a 'science park' as early as 1951, it and other major educational establishments in the area were not at the outset already fully geared-up to produce the research and manpower needs of the new industry; on the contrary, they were at best only potentially capable of meeting these needs as and when they concretely appeared. The local universities and research centres tended to develop *pari passu* along with the industry as a whole. The point here is that wherever the industry may finally have settled down, we would expect to find local educational establishments adjusting their curricula and programmes accordingly.[3]

If Shockley's initial decision and the proximity of major universities to Santa Clara County are really only contingencies, what then constitutes the internal necessary conditions for the consolidation of the industry in central California? I suggest that the development of Silicon Valley can most effectively be comprehended as a process of the reproduction, growth, and transformation of a localised territorial complex of productive labour and social activity.[4] By this I mean that there are determinate dynamics of capitalist industrial and territorial development which ensure that any industry whose market is growing with great rapidity will tend to become rooted ever more insistently in a limited number of areas. There are two major reasons for this. One is that rapid horizontal and vertical disaggregation of production (especially during the early stages of growth) lead to marked agglomeration economies in specific regions. The other is that social and spatial reproduction of the labour force is facilitated by the process of territorial concentration. As these dynamics proceed, an initial locational decision that may well have been nothing more than a caprice, may turn out to be the nucleus of a steadily self-confirming focus of specialized production (cf. Scott 1983). We now need to look at these dynamics in considerably greater detail.

The development of the Silicon Valley production complex

The first thing to note about the initial development of Silicon

Valley, from the late 1950s to the early 1970s, is the extraordinary rate of horizontal disaggregation of the semiconductor industry. Presumably this horizontal disaggregation occurred as a result of limited internal economies of scale in individual production units in the context of an overall market that was expanding at an average rate of over 15 per cent per annum in real terms (see Table 3.3).

Table 3.3 US domestic semiconductor shipments by major product class, 1960–70 (millions of dollars at current prices)

Year	*Diodes, rectifiers and related devices*	*Transistors*	*Integrated circuits*	*Total US domestic shipments*
1960	228	314	29	571
1961	249	316	38	603
1962	268	303	67	638
1963	282	312	190	784
1964	312	323	288	923
1965	379	426	317	1122
1966	471	504	492	1467
1967	444	434	505	1383
1968	420	427	568	1415
1969	464	472	751	1687
1970	421	411	888	1720

Source: United States Department of Commerce (1979: 39).

The outward form of this horizontal disaggregation was a series of 'spin-offs' from pre-existing firms. Thus, as Saxenian (1981) points out, Fairchild Semiconductor was the progenitor of no fewer than fifty companies in the Silicon Valley area that were spin-offs in the two decades between 1959 and 1979. These proliferating companies constituted the core of the developing production complex of Silicon Valley. They were locked deeply into competition with one another and they engaged in advanced forms of product differentiation.

As all of this was occurring, vertical disaggregation was also proceeding apace. In conformity with Adam Smith's celebrated dictum (with its specific analytical shape as defined by Stigler 1951), that the division of labour is limited by the extent of the market, the rapidly rising demand for semiconductors led steadily to a marked breakdown of production activities into many specialized units. Thus a wide variety of input services and subcontracting activities now made their appearance within the Silicon Valley complex. A recent business directory indicates that the following direct inputs (among others) are all currently being produced in the Silicon Valley area: automatic production machinery, testing equipment, measuring devices, high vacuum equipment, encapsulation materials, bonding materials, silicon crystals and wafers, photomasks, epitaxial systems, metal plating, deposition and etching services, circuit designs, R & D services, and so on. This marked vertical disaggregation of the entire production complex served at once to reduce the risks associated with capital investment in a highly competitive market, and (by externalising and pooling many individual demands) to lower the fixed costs that producers had to face.

The overall result of this marked horizontal and vertical breakdown of production activities was the development over the 1950s and 1960s of an intricate system of specialised and ever-changing transactions between producing units. Producers were (and are still today) locked into a labyrinth of costly material flows and face-to-face contacts, and the sheer weight of the spatial costs of these interactions has encouraged plants of all kinds to converge towards one another around their own centre of gravity. The pressures on producers to cluster spatially were all the more irresistible in the earlier period of the development of Silicon Valley when production technologies were extremely unstable and susceptible to brusque changes with correspondingly rapid variations in linkage structures. In these ways, the complex grew and local industrial land-use became increasingly more dense. The industry's highly localised character has been underpinned both by the development of specialized banking services and venture capital firms, and by the many downstream consumers of semiconductors in the San Francisco Bay Area (communications firms, electronic instruments producers, aerospace, computer manufacturers etc.). Further consolidation of the complex was secured by processes of local reproduction of particular forms of labour power.

Social and spatial reproduction of the labour force

In a previous section (pp. 31–4), I suggested that the development of the semiconductor industry's component labour processes resulted in a polarisation in the skill structure of the labour force and subsequently the emergence of socially and spatially segregated labour markets. Nowhere was this more the case than in Santa Clara County. As the industry grew over the 1960s and 1970s and the demands for its products increased, so the total demand for labour also grew. On the one hand, the industry's demands for unskilled and semiskilled labour was largely filled by immigrant female Latino and

Table 3.4 Trends in electronics employment in the San Jose Standard Metropolitan Statistical Area,* 1966–78

Year	A *All employees*	B *Blue-collar employees*	C *Women in blue-collar work*	D *Racial minorities in blue-collar work*	*B as % of A*	*C as % of B*	*D as % of B*
1966	15,317	7,495	5,272	1,731	48.9	70.3	23.1
1969	19,758	6,232	3,701	2,516	31.5	59.4	24.3
1970	25,610	7,960	5,498	2,533	31.1	69.1	31.8
1971	24,504	7,493	5,216	2,142	30.6	69.6	28.6
1972	24,601	8,015	5,925	2,487	32.6	73.9	31.0
1973	33,420	12,813	9,640	4,326	38.3	75.2	33.8
1974	38,122	16,175	12,488	5,523	42.4	77.2	34.1
1975	39,852	13,072	9,520	4,552	32.8	72.8	34.8
1978	41,088	14,391	10,221	6,444	35.0	71.0	44.8

Source: Snow (1982: 43–5).
*Employment principally in Silicon Valley

Asian (especially Filipino) workers (Tables 3.4 and 3.5) who resided in the San Jose area of the County, some distance from the centre of production in such north-County cities as Palo Alto, Mountain View and Sunnyvale. On the other hand, the industry's demands for highly trained scientists, engineers, and technicians have been filled largely by white male graduates of local universities and colleges (Table 3.5), who tend to reside in relatively close proximity to the semiconductor plants and laboratories (Saxenian 1981).

Table 3.5 Santa Clara County high technology employment* by race and sex, 1980

Occupational category	*Size %*	*Sex %*		*Race %*				
		Male	*Female*	*White*	*Black*	*Hispanic*	*Asian*	*Am. Indian*
Managers	14	85	15	88	2	4	5	<1
Professionals	20	82	19	83	2	3	12	<1
Technicians	15	75	25	71	3	10	15	<1
Sales	2	67	33	91	2	3	3	<1
Clerical	15	19	81	77	6	10	6	<1
Craft	7	56	44	63	6	17	14	1
Operatives	24	31	69	49	9	23	19	1
Labourers	2	38	61	41	8	34	17	1
Services	1	86	14	49	12	26	13	1
Totals	100	57	43	70	5	12	12	1

Source: Siegel and Borock (1982, Table 9: 46).
*Predominantly in Electronics

One of the consequences of the structure and social composition of the semiconductor labour force, has been the relative absence of trade union organisation. Of American semiconductor workers who are unionised, 96 per cent are employed in the Northeast of the country, by large, integrated multinationals such as AT & T and ITT. As of early 1987 none of the workers employed by the 'merchant' semiconductor producers in Silicon Valley were unionised.[5] Partly as a result of the lack of unionisation, wage rates for production workers in the Valley were in 1977 between 31 per cent and 61 per cent lower than for their unionised colleagues in the Northeast (Troutman 1980).

During the 1970s and 1980s the semiconductor labour markets, if anything, became increasingly socially and spatially segregated. The production workforce has continued to be reproduced by drawing in workers of Third World origin, a growing proportion of whom are illegal immigrants (Snow 1982; Katz and Kemnitzer 1982). While the social reproduction of this part of the labour force has not thus far produced significant problems for the semiconductor companies, the spatial dynamics of their reproduction has. Specifically, these workers have been subject to rising costs in terms of time and money as a result of having to commute from the relatively distant, but low cost housing areas of San Jose and adjacent communities. It is these problems of commutation rather than the routinised work, or poor wages *per se*, which seem to have led to shortages of unskilled labour, and high turnover rates in the Valley's semiconductor labour force (Saxenian 1981).

The main bottleneck on the further development of the industry in Silicon Valley, in fact, has been the reproduction and supply of highly skilled engineering and technical labour. As indicated above, Stanford University in particular accommodated itself from the start to the basic manpower needs of the industry. Stanford has produced a constant (but always numerically inadequate) stream of graduates at every level of academic attainment from B.Sc. to Ph.D. who have moved directly into jobs in the semiconductor industry. As I have suggested, this phenomenon can scarcely be seen as an 'independent variable' but rather as a subjacent moment of the entire developmental process of Silicon Valley in general. There is a widespread tendency for local colleges and universities to adapt their educational programmes to the peculiarities of local industrial activity, and Silicon Valley (and, in fact, Scotland; see Chapter 6) has been no exception in this regard. Moreover, by acting in this way, they help to socialise the costs of specialized manpower training thereby lowering manufacturers' overheads. The skills of the labour force are finely honed in the work-place, and a many-sided local labour market comes into being alongside the dominant production complex. Firms, then, are able to fill vacancies as they arise from within this pooled labour resource.

At the same time, the labour force must be housed and appropriate forms of communal and recreational activity allowed to develop. It is especially important for the continued viability of the complex to ensure that the delicate norms of neighbourhood activity and environmental quality necessary for the social reproduction of the upper echelons of the labour force be secured. This imperative has been all the more pressing in Silicon Valley in view of the circumstances that upwards of 40 per cent of all workers in the industry have typically been highly educated and highly paid white-collar workers. Saxenian (1981) has described the communities inhabited by these workers as typified by low-density suburban tracts with expensive housing and a wide variety of recreational opportunities. This kind of development is especially characteristic of the affluent northern and western foothill cities of Santa Clara County where a spacious and semi-rural ambience prevails. Local municipalities and planning agencies have helped to underpin this state of affairs by imposing appropriate zoning provisions and density controls. In these ways, then, the form and substance of local urbanisation processes have also helped to cement the complex into a functioning, viable whole.

None of these processes of growth, development, and reproduction proceeds unproblematically, however. Sooner or later various limits to further expansion of the complex begin to make their appearance.

The predicaments of concentrated territorial development

We have seen that industrial complexes tend to grow, in the first instance, at least, via the dynamics of horizontal and vertical disaggregation of labour processes and via appropriate social and spatial reproduction of their associated labour forces. These dynamics tend to produce a situation in which costs of production on the terrain of any given complex fall steadily in relation to all other possible locations, and this then propels the complex forward into yet more advanced stages of development. As this occurs, the complex begins increasingly to encounter internally generated barriers to its own further growth, and problems mount as more and more productive activities and population pile up in the local area. In Silicon Valley, these problems started to become especially pronounced after the late 1960s; shortages of engineering and technical labour became chronic, wages of all sections of the labour force escalated upwards, land values increased, and a severe shortfall of adequate housing seemed to become endemic to the whole area (Bernstein, *et al.* 1977: Saxenian, 1981). In view of these difficulties, it was not long before representatives of the semiconductor industry were seeking ways of emancipating themselves from that which they had created at the outset. They did this by technical and organisational restructuring and dispersing new investment in particular component labour processes to profitable production locations elsewhere.

Before we continue with our account of the spatial development of the industry, we need to return to our examination of the technical and social processes underlying this development.

Capital deepening, markets, and the determinants of globalisation

From its very inception, the semiconductor industry has tended to become increasingly capital-intensive. The cost of setting up a state-of-the-art semiconductor manufacturing plant in 1965 was only about $1 million; by 1980, this cost had escalated to $50 million (Saxenian 1981). Simultaneously, the character of the industry was changing. This was reflected in part in the growing proportion of integrated circuits (as opposed to discrete devices) in total semiconductor output, as it was (in part) in the increasing tendency to standardisation of outputs in important market segments.

In the early years of the industry, many relatively small firms emerged in Santa Clara County around an insistent process of horizontal and vertical disaggregation in which, in addition, firms provided more or less customised products in response to specific market 'niches'. As I have already demonstrated, the result of this process was not only a mushrooming of producers in Santa Clara County, but also the development of an industrial system based on small-batch production. In this kind of economic environment firms were limited in the levels of automation that they could achieve. In particular, it was important for them to be able to switch easily from one product line to another, and this necessitated relatively low capital-labour ratios (Keller 1981). Further, in the early years, the number of bonding operations to be carried out per device in the assembly stage was small, and so manual labour was not yet the barrier to profitability and expansion that it was later to become. In fact, this barrier only really became decisive in the late 1970s, and automated assembly procedures began to be assimilated on a large scale into production activities in order to counteract it.

Even so, from the early 1960s, tendencies to technical change, capital deepening, and the standardisation of product lines (in the latter case, particularly with regard to discrete devices) were apparent. At about the same time, American producers had begun to come under intense pressure from Japanese competitors who were able to utilize their own cheap labour and develop their production capacities on the basis of state protection for the domestic market (see Table 3.6 for an indication of the strength of Japanese competition in major transistor markets). What is more, during that same period there began to be a relative decline in military purchases of semiconductors and a relative growth in civilian markets where product demands were less specialized. In 1960, military end-users still consumed 50 per cent of US semiconductor production, but by 1966, this had dropped to 30 per cent, and then by 1972 to 24 per cent (Braun and Macdonald 1982). This relative decline in military markets had three major corollaries. First, as noted, it was associated with an increased relative (and absolute) level of large-scale, standardised demands. Second, it induced stronger price-competition and the necessity for more stringent cost-cutting measures (since reliability more than cost is the concern of military purchasers). Third, it opened the way to the internationalisation of production, for military procurements, by federal law, and almost entirely restricted to domestic manufacture (Snow 1982). All of this has encouraged a series of significant locational shifts in the industry.

Table 3.6 US and Japanese transistor output during the transition to overseas production, 1957–68 (output in millions of units)

	Output		*Percentage of*	*Percentage of*
Year	*United States*	*Japan*	*Japanese transistors used in radios*	*transistor radios exported*
1957	29	6	67	n.a.
1958	47	27	67	n.a.
1959	82	87	55	77
1960	128	140	48	70
1961	191	180	41	67
1962	240	232	34	76
1963	300	268	35	81
1964	407	416	33	69
1965	608	454	30	75
1966	856	617	26	86
1967	760	766	23	83
1968	883	939	20	90

Source: Grunwald and Flamm (1985, Table 3.4: 70).

Determinants of the globalisation of the American semiconductor industry

I have already indicated that the production of semiconductors involves five technically disarticulated labour processes, i.e. R & D, mask making, wafer fabrication, assembly, and testing. The different labour processes in the industry have widely varying needs in terms of capital investment, labour skills, specialized inputs, and so on. Although these component labour processes were in general all found in geographical association with one another in the early years of the industry's growth in Santa Clara County, the fact that they were technically disarticulated meant that they could be organisationally and hence spatially separated. It is true, there are definite internal economies of scope and scale that tend to keep all the labour processes in the semiconductor industry unified within the single firm. Nevertheless, if linkage costs on their transactional activities with one another are sufficiently low, then the possibility of their spatial dispersal can be made real. Cheap air transport and telecommunications have potentiated exactly this outcome.

That said, there are other strong determinants of the location process in the semiconductor industry. The precise ways in which these determinants play themselves out depend very much on which type of labour process we are talking about. R & D requires ready access to highly trained scientists and engineers for its successful operation; as a corollary, it seems to need the kind of local urban/environmental conditions (such as are found in and around Silicon Valley) which can sustain the effective social reproduction of this form of labour; and, of course, the localised training of such labour in universities and research institutes is a further major asset. Many of the same conditions apply to the location of mask-making and wafer-fabrication facilities. However, there has also been a moderate internationalisation of these facilities, particularly wafer fabrication, in part in order to evade certain kinds of tariff barriers erected by certain national or regional states and to penetrate increasingly lucrative markets. Thus companies such as Motorola, National Semiconductor, Hughes and General Instrument have located water fabrication plants in Britain as a means of successfully penetrating the high (17 per cent) EEC tariff barrier on semiconductors and of therefore tapping the European market (see Chapter 6). Assembly is typically an unskilled operation and it has been increasingly relocated out to the periphery of the world-system over the last two decades. Though, as we shall see (in Chapters 4 and 5), testing functions have in recent years been dispersed to selected peripheral locations also, and most recently there has been limited evidence of the emergence of wafer fabrication plants.

All component labour processes were effectively confined to the United States in the early period of development when markets were limited and specialised, when they were monopolised by US producers and when the military constituted the principal end-user. With the rise of more standardised commercial markets (at first overwhelmingly associated with demands for transistors for the booming radio industry), and the emergence of intense Japanese competition (Table 3.6), the conditions for the internationalisation of particular labour processes were brought to fruition.

Figure 3.2 depicts the dispersal of the labour processes of selected US semiconductors producers, as they appeared in the mid- to late 1970s. A number of important features of the topography of dispersal are evident in Figure 3.2. Observe firstly the almost total concentration of corporate control, R & D, mask making, wafer fabrication and testing in the United States. Observe also, however, the emergence of wafer fabrication, design centres (engaging not in

Figure 3.2 International division of labour in US semiconductor production: Selected companies and locations, *circa* 1976–9

Company	*Nature of operation and labour processes*														
	USA	Scotland	England & Wales	Germany	France	Switzer-land	Japan	S. Korea	Taiwan	Hong Kong	Singapore	Malaysia	Philippines	Thailand	Indonesia
Motorola	c, rd, w, t, ms, m	w, t		w, t	w, t	d, r, m	w, d	a	a	a, m		a	a		
National Semi-conductor	c, rd, w, t ms, m	w, d	r, m							a, m	a, m	a	a	a	a
Fairchild	c, rd, w, t, ms, m		d, r, m	w	w, t m		w, d	a		a, m	a		a		a
Texas Instru-ments	c, rd, w, t, ms, m				w, t, r		w, d		a		a, m	a	a		
General Instru-ment	c, rd, w, t, ms, m	w, d	d, m, r		m				a			a			
Hughes	c, rd w, a t, ms, m	w, t, (ms)	d, r, m							(a)			(a)		

Siliconix	c, rd, w, t, ms, m	a, t, m	a	a			
Teledyne	c, rd, w, t, ms, m			a	a		
Advanced Micro Devices	c, rd, w, t, ms, m				a	a	a
Silicon Systems	c, rd, w, t, ms, m			a	a		
Sprague	c, rd, w, t, ms, m		a				a
Zilog	c, rd, w, t ms, m					a	

Key: c – corporate control; w – water fabrication; r – regional headquarters; rd – research & development; a – assembly; m – marketing centre; d – design centre; t – final testing; ms – mask-making; () – operation under sub-contract arrangement

Sources: Interviews with company executives; company reports; trade press; Siegel, 1980; Scott, 1987; UNCTC, 1986; Hong Kong Productivity Centre *Electronics Bulletin* (various issues); Scottish Education and Action for Development (1984).

innovative work, but adapting standard designs to customer requirements) and testing functions in certain European locations (in the case of wafer fabrication, particularly, Scotland) and Japan. In both of these cases, the development of such facilities was almost certainly in response to growing markets in those countries, both protected as they were (and still are) by high tariff barriers. Finally, note that assembly functions had been overwhelmingly relocated to peripheral locations in East Asia; (and to a lesser extent Latin America and the Caribbean). By the mid- to late 1970s, no other labour processes other than assembly had been implanted by US companies in that region.

The reasons for this particular international division of labour, the way it has changed in recent years, and its social, spatial, and developmental consequences for some of the countries indicated in Figure 3.2 will be explored in the next three chapters. We begin, first of all, with semiconductor production in the developing countries of East Asia.

Chapter four

East Asia: The emergent regional division of labour

In this and the succeeding chapter, we turn to examine the origins and evolution of US semiconductor production in the developing countries of East Asia. In the current chapter, I argue that while initially US manufacturers may have invested in production facilities in East Asia in order to reduce costs by applying the region's huge supplies of cheap labour to production, the subsequent development of the industry there cannot be understood in terms of such a narrow, unicausal explanation. Specifically, I suggest that a distinct regional division of labour has now emerged with its own (albeit crudely defined) 'cores' and 'peripheries'. The emergence of these cores has not been associated with their supplies of cheap manual labour, but if anything, with their increasing ability to provide good quality engineering and technical labour, with the development (in some, though not all cases) of their own semiconductor production complexes, and finally, with particular forms of state intervention.

In this chapter, I describe the evolving topography of the East Asian regional division of labour, and advance an explanation for its development. In the next chapter, I examine in more detail the emergence of one of the regional cores of semiconductor production: Hong Kong. The likely future evolution of the regional division of labour, and the developmental prospects for the various territorial units which compose it, will be dealt with in the penultimate chapter.

Globalisation of production

I argued in the previous chapter that US semiconductor companies, confronted by stiff Japanese competition in discrete (especially transistor) markets, internationalised their most labour-intensive production process (assembly) as a cost-lowering strategy (Okimoto *et al.* 1984; Grunwald and Flamm 1985; UNCTC 1986). It was a relatively 'mature' semiconductor technology, then, that was first internationalised. Furthermore, internationalisation, in the 1960s and through to the present day, has been a cost-lowering strategy utilized primarily not simply by American, but more specifically by American 'merchant' semiconductor manufacturers.[1] American electronics systems corporations (such as IBM and AT & T) which control their own semiconductor divisions, and manufacture largely for their own consumption, have only rarely set up semiconductor plants (assembly *or* wafer fabrication) outside of the United States.

In addition, it was predominantly the merchant manufacturers of standardised, large-volume semiconductors who initially developed plants in East Asia (Motorola, Fairchild, National Semiconductor etc.), rather than those whose major business is in customised, low-volume devices (Hughes, Teledyne, Zilog etc.). As we shall see in a later chapter, the growth of customised *vis-à-vis* standardised markets in recent years, is one of the developments which may have significant implications for the future topography of the East Asian regional division of labour.

Just as Fairchild Semiconductor had been the originator of the Silicon Valley electronics complex in the late 1950s, so it initiated offshore semiconductor production in East Asia in the early 1960s. Hong Kong was the first East Asian recipient of Fairchild investment in 1961 when the firm established a plant to assemble transistors and subsequently integrated circuits. From this beginning in Hong Kong the industry diffused out to other locations in East Asia, and this diffusion eventually brought in its train a regional division of labour. Before analysing the structure and significance of this regional division of labour, however, we need briefly to attend to the reasons why Hong Kong should have been the initial preferred location for US semiconductor investment. (We will address these issues also, but in much more detail, in the next chapter).

In Chapters two and three it was argued that the determinants of the globalisation of industrial production at the highest level of abstraction arise out of the need for capital valorisation and accumulation. Empirically these determinants have associated with them spatially and historically contingent circumstances (e.g. local reproduction of particular kinds of labour power, or as in Silicon Valley, the development of restrictive planning legislation) which tend to produce particular locational outcomes for industrial investment. Although, as I have suggested, problems of realisation, and hence the need to penetrate particular, protected markets sometimes operate as important locational determinants, this was not especially the case with regard to the emergence of semiconductor production in East Asia. From initial internationalisation through to the present day the United States constituted, overwhelmingly, the principal market for US firms producing in the region (UNCTC 1986).

In the case of semiconductor production in East Asia the most important initial determinant was, as I have already argued, the presence of enormous pools of cheap and underemployed labour (Rada 1982). But the mere existence of cheap labour was in itself not a sufficient locational inducement. Hong Kong had a number of additional special advantages, which made it a particularly attractive location. (Similar advantages in Taiwan, South Korea, and Singapore meant that they were also to develop flourishing semiconductor industries at an early stage). These advantages included (a) political stability, (b) an open financial system with no limits on the repatriation of profits, and (c) excellent telecommunications and air transport facilities (Henderson and Cohen 1982a). In addition, Hong Kong had a further crucial advantage. Over the 1950s, it had developed a flourishing industrial economy based on textiles, garments, plastics, and other labour-intensive forms of production. By the late 1950s, Hong Kong was already integrated into the NIDL and had become a major location for radio assembly and production of the cheaper varieties of consumer electronics (Chen 1971). This meant that Hong Kong possessed by the early 1960s a work force that was habituated to the kinds of labour processes characteristic of semiconductor assembly. It had, in other words, what many observers (including manufacturers) wrongly classify as a 'skilled' labour force. Furthermore, the existence of a flourishing informal sector in Hong Kong, together with state subsidies, helped to keep industrial wages down to a level which at that time was quite comparable to wages in many other peripheral areas. Finally, Hong Kong was able to supply the small but crucial demand for qualified engineers and technicians necessary for the successful operation of semiconductor assembly processes.

By investing in a production facility in Hong Kong, Fairchild established an international division of labour in semiconductor manufacture which, though it was altered in form, effectively set the pattern for the subsequent expansion of offshore production. Thus Taiwan was opened up by General Instrument in 1965 and South Korea by Fairchild and Motorola in 1966. By 1968, Texas Instruments, National Semiconductor, and Fairchild had set up plants in Singapore. 1971 saw the emergence of semiconductor production (initially by National Semiconductor) in Malaysia, while by the mid-1970s, the Philippines, Thailand, and Indonesia had also been incorporated into the division of labour.

By the mid-1970s, then, US semiconductor plants had been established in every capitalist East Asian developing society other than Brunei, and by the mid-1980s, particularly heavy concentrations (in terms of number and size of plants) had emerged around Penang and

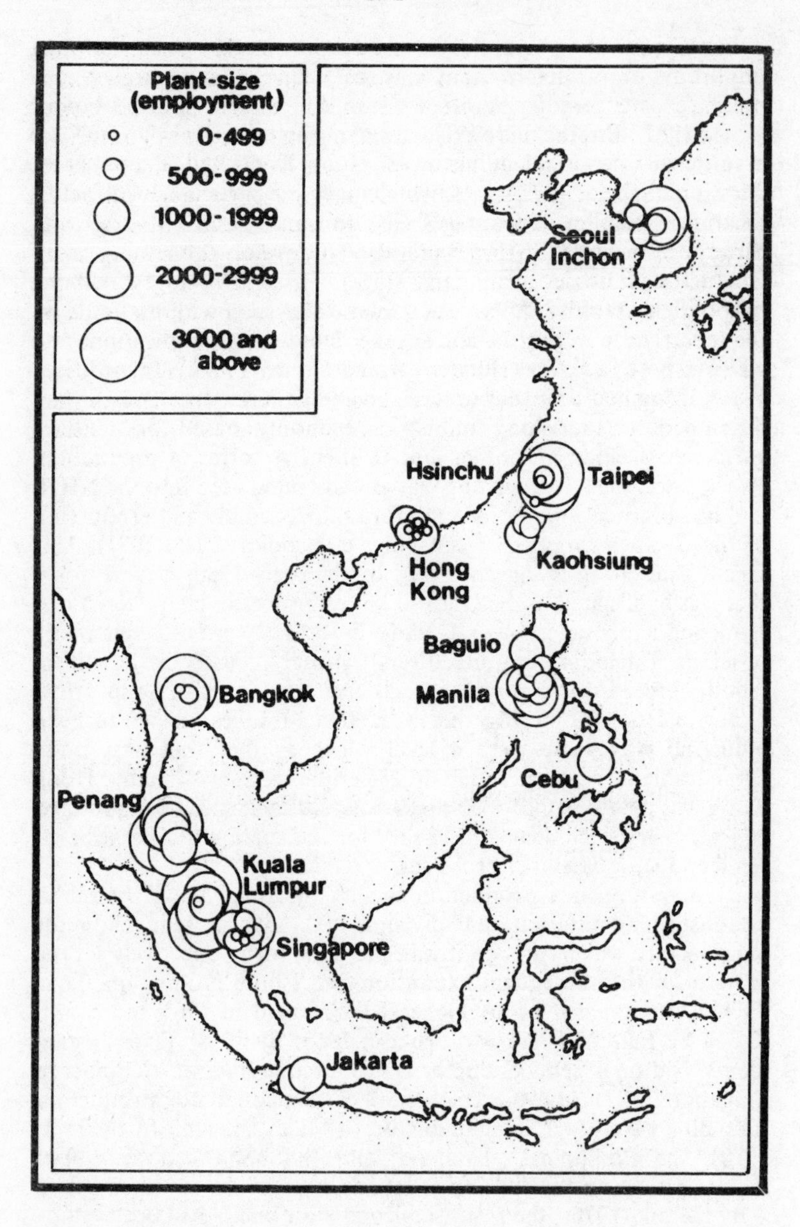

Figure 4.1 US-owned semiconductor plants in the Developing Countries of East Asia, 1985

Source: Scott (1987: 149)

Kuala Lumpur in Malaysia, Manila in the Philippines, and Singapore (Figure 4.1). What is more important for our immediate concerns, however, is not so much the geographical spread and concentration of US semiconductor production in the region, but rather the roles the various plants have played within the industry's international division of labour, and how those roles have changed over time. Our attention, then, is directed to the particularities of the industry's labour processes and managerial functions and the nature of their dispersal across the region.

By the mid- to late 1970s, the semiconductor division of labour appeared (for a selection of US companies) as depicted in Figure 3.1 (p. 32). In addition to the general comments I have already made (in Chapter 3) about the data summarised in Figure 3.1, there are a number of additional comments, specifically about East Asia, that need to be recorded.

1. There was an exlusive concentration not only of organisational control functions in the USA (as one would expect), but also of the labour processes requiring the most advanced scientific and technological skills (R & D), and to a large, though lesser extent, those containing the highest levels of capital intensity (mask making and wafer fabrication). Note also the high preponderance of testing functions, as well as limited assembly functions, in the latter case associated largely with the production of small batch, customised semiconductors.
2. With the exception of those territorial units that had marketing responsibilities (principally, at this time, Hong Kong and Singapore), only the most labour-intensive assembly processes had been implanted in the region. In the case of each territorial unit, penetration of the domestic economy seems to have resulted largely from a combination of three factors: the need for expanded (and in some cases duplicate)[2] assembly capacity in the context of rapidly increasing demand; the search for ever-cheaper supplies of unskilled labour once wage rates in the initial recipients of investment – Hong Kong, Taiwan, South Korea, and Singapore – began to rise (see Table 4.1); and policy initiatives by the national states themselves (export processing zones, Silicon Valley 'shopping expeditions' as in the case of the Malaysian Government, etc.). This initial spread of US semiconductor production into East Asia does indeed seem to have been associated, as Fröbel *et al.* (1980) suggest and 'early' studies of the experience of the predominantly female workforce confirm (for instance, Djao 1976; Lim 1978b; Grossman 1979), with the search for new opportunities for the creation of absolute surplus value.

Table 4.1 Hourly compensation of production workers in the East Asian electronics industry: Relative and absolute values

	1969[1] *(US = 100)*	*1975*[1] *(US = 100)*	*1985*[2,3] *(US = 100)*	*1985*[2] *Average hourly Wage (US$)*
Hong Kong	10	12	16	1.33
Indonesia	n.a.	5	4	0.35
Korea	10	7	14	1.19
Malaysia	n.a.	9–10	10	0.84
Philippines	n.a.	6	8	0.63
Singapore	9	12	19	1.58
Taiwan	8	7	16	1.36
Thailand	n.a.	5	5	0.43

Source: Scott (1987, Table 1: 145).

Notes: 1. Compensation in the international electronics industry (Grunwald and Flamm 1985, Table 3.9).
2. Compensation in US-owned and locally owned semiconductor assembly plants (Scott's questionnaire survey).
3. US standard of comparison given as $8.37, i.e. average hourly earnings of production workers in SIC 367 (Electronic Components and Accessories) for August 1985 (US Department of Labor, Bureau of Labor Statistics, Employment and Earnings).

3. All the firms assembling semiconductors in the region at this time organised their divisions of labour in more or less the same way. The wafers were fabricated in the United States, air-freighted to East Asia, assembled into discretes or integrated circuits, and air-freighted back to the United States for final testing. As I have already indicated, the vast majority of the output was destined for American markets, though what was not, was in any case marketed by the parent companies (Siegel 1980). The emergence of this spatial division of labour was effectively encouraged by the US Government's tariff regulations 806.30 and 807.00 which charged import duty only on the value added abroad. As the value added to semiconductors assembled offshore was primarily a result of the application of cheap, unskilled labour power, duty charged on their re-import did not substantially affect the economic viability of offshore production.
4. Another feature of this division of labour, was the form which the investment took. Fairchild and every American semiconductor manufacturer who followed them to East Asia chose to establish wholly owned subsidiaries in the region, rather than joint-venture or licensing operations. The reasons for firms' preference for this form of production seem clear enough. Reliable production methods and high yields (in circumstances where both produc-

tion technologies and product specification were changing rapidly) depend on rigorous technical and managerial control. Even at the present time, offshore production of semiconductors is overwhelmingly conducted in wholly owned branch plants.[3] There has, however, been a significant development of independent firms (particularly in the Philippines and South Korea) who assemble under subcontract arrangements, particularly for smaller producers, or at times when the larger manufacturers are unable to cope themselves with excess demand (Paglaban 1978; UNCTC 1986). We shall discuss subcontract assembly, however, in more detail in a later section.

The regional division of labour

The US semiconductor division of labour as depicted in Figure 3.1, approximates the division of labour in export-oriented, transnationally organised, industrial production that one would expect from reading the work of Frank (e.g. 1981), Fröbel, *et al.* (1980), and others who work in a similar vein. Whatever the topography of semiconductor labour processes and the relations of dependency that have held them together across the globe, however, recent developments in the industry in East Asia, tend to raise doubts about this line of analysis. When we come to examine the determinants of the industry's development in the region (and particularly in Hong Kong), we will see that perhaps more serious doubts must be raised about other, and rather different attempts, to explain the phenomenon of industrialisation and economic growth more generally. Before we reach that point, however, we need to describe and comment upon the ways in which the semiconductor division of labour has changed in recent years.

Figure 4.2 represents the US semiconductor division of labour as it had evolved by the mid-1980s. We notice immediately that certain changes seem to have taken place. Most importantly for our current purposes, a specifically East Asian division of labour has developed. This division of labour consists of the following elements:

1. The 'gang of four', but especially Hong Kong and Singapore have become the recipients of investment in final testing facilities. As I indicated in the previous chapter, testing has become a largely capital-intensive labour process (involving in its most advanced forms, lasers coupled to computers), and therefore, as a result, requires significant numbers of highly qualified engineers and technicians. Furthermore, in the case of Motorola, Fairchild and Sprague in Hong Kong, and National Semiconductor and

Figure 4.2 International division of labour in US semiconductor production: Selected companies and locations, *circa* 1985–8

Company	*Nature of operation and labour processes*															
	USA	Scotland	England & Wales	Germany	France	Ireland	Switzer-land	Japan	S. Korea	Taiwan	Hong Kong	Singapore	Malaysia	Philip-pines	Thai-land	Indonesia
Motorola	c, rd w, a, t, ms, m	w, a t		w, t	w, t		d, r, m	w, d	a	a	d, t r, m		a	a		
National Semiconductor	c, rd, w, a, t, ms, m	d, w	r, m								a, t, m	t, d, m, r	w, a, t	a	a, t	a
Fairchild	c, rd, w, a, t, ms, m		d, r, m		w, t, m			w, d, a, t	a, t	d	t, m	a, t, m		a		a
Texas Instruments	c, rd, w, t, ms, m				w, t, r			w, d		a		a, t, m	a	a		
General Instrument	c, rd, w, t, ms, m	w	d, m, r		m					a, t			a			
Hughes	c, rd, w, a, t, ms, m	w, a, t (ms)	d, r, m								(a)			(a)		

Siliconix	c, rd, w, t, ms, m		a, t, m			a	a, t, d				
Teledyne	c, rd, w, t, ms, m						a, t				
Advanced Micro Devices	c, rd, w, t, ms, m			w			m	t, m	a	a	a
Silicon Systems	c, rd, w, t, ms, m							d, a, t			
Sprague	c, rd, w, t, ms, m					a	r, t, m			a	
Zilog	c, rd, w, t, ms, m			w			d, r, m			a, t	
Burr-Brown	c, rd, w, t, ms, m, a	w, a, t, d			a, t						

Key: c – corporate control; w – wafer fabrication; r – regional headquarters; rd – research & development; a – assembly; m – marketing centre; d – design centre; t –final testing; ms – mask-making; () – operation under sub-contract arrangement

Sources: Interviews with company executives; company reports; trade press; Henderson (1987); Scott (1987); Ehrlich (1988); Hong Kong Productivity Centre, *Electronics Bulletin* (various issues).

Advanced Micro Devices (AMD) in Singapore, assembly facilities have been eliminated.[4]

2. Circuit-design centres have also been set up in Hong Kong and Singapore by a number of US companies. Motorola, Siliconix, and Zilog all have design centres in Hong Kong,[5] while National Semiconductor and Silicon Systems have theirs in Singapore. Fairchild has recently opened a design centre in Taiwan. It must be recognised, however, that none of these design centres perform innovative R & D (such R & D is still done by US companies, only in the United States), but rather adapt basic designs to customer requirements. Even so, they still require skilled engineering personnel.
3. Alongside the specialisation of Hong Kong and Singapore in testing and design, a number of companies have established their Asian regional headquarters there. Motorola, Sprague, and Zilog have done so in Hong Kong, as have National Semiconductor in Singapore. Generally speaking, the developments adumbrated above seem to indicate a shift from concerns with absolute, to relative surplus-value production in the 'cores' of the regional division of labour.
4. In 1987 National Semiconductor became the first US company (and only the second foreign company,[6]) to establish a wafer fabrication plant in the developing world when it began fabrication at its plant in Penang, Malaysia. Although the plant does not fabricate the entire wafer, but rather performs one small part of the process, other US companies such as Intel and Motorola have plans to develop full wafer-fabrication capabilities in their Malaysian plants (Seaward 1987; Ehrlich 1988). While these developments in Malaysia are of considerable significance, it should be borne in mind that the design centres for these companies, and their managerial control functions, are located elsewhere in the region, particularly in Singapore and Hong Kong. We shall return to these issues later in the chapter.
5. In spite of the very recent developments indicated above, Malaysia together with the other territorial units in the regional division of labour (Thailand, the Philippines, Indonesia) continue to concentrate on assembly. However, here and there, testing facilities have been installed (for instance, by National Semiconductor in Malaysia and Thailand), but it is probably the case that it is the technologically simpler devices (discretes and less-sophisticated memory ICs) that are tested there, rather than, say, in National Semiconductor's principal testing centre in Singapore. Furthermore, in some cases the automation of assembly processes has taken place. This has been particularly

true in Malaysia, though to a far lesser extent, for instance, in the Philippines (UNCTC 1986; Salih and Young 1987).

6. It appears, then, that it has been the plants on the periphery of the regional division of labour that have become (perhaps) increasingly important for assembly work *vis-à-vis* other locations. Notwithstanding the technological developments indicated in paragraphs 4 and 5, in so far as the average employment size of plants can be taken as a guide (albeit a weak one) to the relative proportion of assembly work performed by the various territorial units within the regional division of labour, then the data presented in Table 4.2 would seem to confirm such a hypothesis.[7]

Table 4.2 US semiconductor plants in East Asia: Numbers and employment size, 1985

Country	*Number of plants*	*Total employment*	*Average size of plants*
Hong Kong	8	4,552	569
Indonesia	2	3,200	1,600
Korea	5	8,800	1,760
Malaysia	14	38,136	2,724
Philippines	11	13,112	1,192
Singapore	11	10,397	945
Taiwan	8	15,296	1,912
Thailand	4	6,470	1,685
Totals	63	99,963	1,587

Source: Adapted from Scott (1987, Table 5: 147).

Before we turn to examine the articulation of the regional with the overall international division of labour within the industry, and to assess the role of the state and the emergence of production complexes etc. for the development of the industry in the various territorial units, it may be useful to ask why this peculiar regional division of labour has occurred. Why, in other words, has more and more of the investment in assembly plants for large-batch standardised outputs tended to go to an increasing degree to Thailand, Malaysia, the Philippines, and so on, while Hong Kong, Singapore and now, seemingly Malaysia, have tended to be upgraded as to the quality and complexity of their production technologies and labour processes? There are two major reasons for this. First, as the data in Table 4.1 indicate, during the 1970s and 1980s, manual labour costs, not simply in Hong Kong and Singapore, but in Taiwan and South

Korea also, increased substantially relative to those in the units at the periphery of the regional divison of labour. Thus much of the new investment in labour-intensive assembly work (given reasonable levels of political security – something which has been particularly a problem in the Philippines in recent years) has shifted to these persistently low-wage countries. Second, Hong Kong and Singapore (and more recently South Korea and Taiwan) have long been able to supply well-trained technicians, engineers, and scientists. This is a reflection, in part, of their more advanced level of development, as it is also, in part, of their advanced educational systems that can generate large numbers of highly qualified personnel. Thus, critical testing and design activities can be carried on in these centres with quite high levels of reliability as well as at a relatively cheap price *vis-à-vis* the USA, Japan or the EEC (see Chapter 5 for details on the Hong Kong case).

The regional division of labour can be seen, to some extent, in value-added data for semiconductor imports into the United States. As 'early' as 1979, the value-added to semiconductors partially processed in Hong Kong and imported into the United States was 56 per cent of their total price. For semiconductors imported from Singapore (and Taiwan), the corresponding figure was 46 per cent. For Malaysia, Korea, and the Philippines, the value-added was, respectively, 45, 39, and 32 per cent of the price (Ernst 1983: 166).

Articulation of the divisions of labour

Whatever the contours of the East Asian division of labour in semiconductor production, and the developmental tendencies inherent within it, there is no possibility of developing an adequate explanation of its significance, unless it is recognised that its features and the way they alter over time are a product of the articulation of the regional with the international division of labour, and of the dynamic development of that articulation.

In spite of the fact that certain territorial units within the regional division of labour have emerged as semiconductor production centres producing a technologically more advanced output, employing greater technical expertise, and exercising greater managerial and marketing control than elsewhere in East Asia, it must be remembered that this 'advanced' development is relative, and appears to have significant limitations placed upon it by virtue of the articulation of the divisions of labour. Let me sketch out some of the features of this articulation as they currently exist, and particularly examine some of the flows and dependencies associated with it.

1. Whereas previously each plant, or national operation within the

region, would assemble its quota of semiconductors and ship them directly to the United States for testing and distribution to domestic and world-wide markets, the emergence of the regional division of labour with its own 'cores' and 'peripheries' has altered both product flows and the spatial chain of command. What increasingly tends to happen is that although fabricated wafers are shipped from the United States or Europe directly to assembly plants in, say, Malaysia or the Philippines, once assembled, they are routed to the respective company's testing centres in Hong Kong and Singapore, and from there directly to US, European, or Asian customers. The data assembled in Table 4.3 provide a flavour of these product flows.

Mark, in particular, the large imports of integrated circuits from South Korea and Malaysia into Hong Kong and from Thailand, Malaysia, and the Philippines into Singapore. While it is likely that some of the Korean imports into Hong Kong are devices destined directly for the territory's watch-assembly industry (the largest in the world), it is almost certainly the case that imports from the other countries mentioned, into both Hong Kong and Singapore, are destined for the respective companies' testing centres in those locations prior to final export. Of additional note in Table 4.3 is the relatively large reverse flow of semiconductors from Singapore to Malaysia. Given that testing facilities have also emerged in Malaysia (Figure 4.2), this data probably suggests that it is the technologically simpler devices that are shipped to Malaysia for testing (for instance by National Semiconductor) whereas more sophisticated ICs are tested in Singapore (see UNCTC 1986: 398–407).

2. While it is important to bear in mind that different corporations have different structures of control throughout their global operations (Motorola, for instance, allowing its plants relative degrees of autonomy; National Semiconductor, on the other hand, centralising its authority at its Silicon Valley headquarters), the emergence of Hong Kong and Singapore as regional headquarter and marketing locations, gives them important leverage over the shape of the regional divison of labour, and greater relative weight at corporate headquarters, in terms of decisions affecting any given location within the firm's Asian-Pacific operations.
3. While technological upgrading and some regional control capacities have gravitated to places such as Hong Kong and Singapore, they and every other territorial unit within the regional division of labour, remain ultimately dependent on corporate headquarters in the United States, totally technologic-

Table 4.3 Hong Kong and Singapore: Imports and exports of integrated circuits (Selected regions/countries), 1981 (Quantity and value: 000s of Units and US $)

Region/Country	*Hong Kong*				*Singapore*			
	[a]Imports		[a,b]Exports		[a]Imports		[a,b]Exports	
	Quantity	Value	Quantity	Value	Quantity	Value	Quantity	Value
North America								
USA	50,848	59,132	76,357	62,308	63,947	38,351	10,293,332	376,472
Canada	960	1,448	1,529	2,670	66	57	8	19
Western Europe								
EEC	10,962	8,120	32,570	30,862	30,075	10,438	405,197	126,647
Other	710	320	6,016	5,089	1,932	1,177	10,133	1,406
Latin America	104	57	63	79	1,502	1,179	356	212
Asia								
South Korea	217,030	66,243	873	833	9,403	2,889	14,205	1,414
Thailand	35,385	24,846	364	123	152,660	107,885	15,835	4,398
Japan	93,811	63,677	898	1,110	70,344	31,515	111,624	27,970
PRC	1	1	975	846	N/A	N/A	N/A	N/A
India	–	–	1,046	N/A	20,159	1,916	1,065	511
Western Malaysia	109,390	21,349	3	3	526,703	114,646	120,070	12,135
Singapore	80,790	29,042	1,257	685	–	–	–	–
Hong Kong	–	–	–	–	107,754	14,255	435,612	50,862
Taiwan	26,371	49,896	10,356	4,915	15,584	6,457	33,611	8,989
Indonesia	21,649	5,150	–	–	N/A	N/A	N/A	N/A
Philippines	13,344	11,471	1,664	932	148,214	30,929	46,828	4,261
Australia	68	297	2,842	1,093	435	68	15,965	4,190
Totals	661,423	341,047	136,813	111,548	1,146,869	361,712	2,239,841	619,513

Source: Compiled from UNCTC (1986, Tables VIII.2, VIII.8: 386, 399).
Notes: (a) Assembled ICs only
(b) Domestic Exports

ally dependent on American R & D and circuit masks, and substantially dependent on wafers fabricated in the US or Europe. It must be emphasized once more, that up to the present time, no American, nor indeed Japanese or European semiconductor house, has invested in R & D, or mask making or wafer fabrication – the three technological kernels of the entire operation – other than in the US, Japan, or the EEC, with the sole exceptions of wafer fabrication in Singapore and Malaysia noted above.

Surveying the strengths, weaknesses, and dependencies within the East Asian division of labour, does not provide us with a full sense of the extent to which the various territorial units within it may be able to develop viable, indigenously controlled semiconductor industries of their own. Nor does it provide us with a sense of how social forces internal to the particular territorial units have helped to produce their current roles within the division of labour.

In the next sections, therefore, we move to examine some of the general economic, social, and political issues surrounding the emergence of particular territorial units as cores or peripheries of the regional division of labour. We begin with the emergence, in a number of locations, of indigenously owned semiconductor production complexes. The development of these complexes combines questions of the role of the national state in economic growth with issues of how foreign manufacturing investment can, under certain circumstances, help to stimulate local industrialisation. As I deal with Hong Kong separately in the next chapter, in what follows we will concentrate on other parts of East Asia.

Growth of semiconductor production complexes

One of the key questions about the internationalisation of any industrial branch (or sector) involves the extent to which plants associated with that branch can help to stimulate industrialisation within the 'host' territorial unit. The emergence of local complexes of productive activity may result, of course, from the attraction of other foreign investment to the locality, and/or it may result from the growth of indigenously owned and controlled enterprises. From the point of view of a country attempting to create a degree of relative autonomy for itself within the world economy, and hence loosen the bounds of dependency, then the extent to which it is able, ultimately, to organise and control its own industrial base, is a matter of some importance.

As we might expect, the extent to which foreign semiconductor investment has been able to stimulate local production complexes varies enormously from place to place within the East Asian division of labour. While, generally speaking, semiconductor production complexes can be seen as emerging in the cores, rather than the peripheries of the regional division of labour, this is not uniformly the case. Thus, for instance, while Singapore has developed as one of the technological and control centres of the US-owned industry in the region, so far, there seems to be a relative absence of locally owned firms performing, for instance, sub-contract assembly work or supplying materials or special services (Lim and Pang 1982). In addition, Singapore, unique amongst the 'gang of four', does not yet have a locally owned semiconductor plant with wafer fabrication or other more technologically advanced capabilities (UNCTC 1986: 389–407). On the other hand, while there is little evidence that foreign investment has stimulated local production complexes in peripheral units such as Malaysia, Thailand, or Indonesia, there is evidence that the Philippines has emerged with a variety of locally owned firms performing subcontract assembly work and other specialised services such as machining, metal plating and stamping, and test and 'burn-in' work (Scott 1987).

Table 4.4 Semiconductor production complexes in the East Asian division of labour: Locally owned plants, 1985

	Integrated Wafer fabrication/Assembly plants			*Sub-contract assembly-only plants*		
Country	Number of plants	Total employment	Average workforce per plant	Number of plants	Total employment	Average workforce per plant
Hong Kong	4	1,532	383	2	2,700	1,350
Indonesia	–	–	–	–	–	–
Korea	5	5,000	1,000	12	10,474	873
Malaysia	–	–	–	2	1,450	725
Philippines	–	–	–	14	18,046	1,289
Singapore	–	–	–	1	240	240
Taiwan	8	5,064	633	11	2,805	255
Thailand	–	–	–	2	900	450
Regional totals	17	11,596	682	44	36,615	832

Source: Adapted from Scott (1987, Table 5: 147).

Table 4.4 gives an indication of the extent and distribution of locally owned semiconductor plants (i.e. those that fabricate wafers as well as perform assembly functions) and subcontract assembly houses. It is immediately clear that both wafer fabrication and assembly plants are overwhelmingly located in those territorial units that have emerged as the cores of the regional division of labour. The exceptions, as already noted, are the relative absence of locally owned plants in Singapore, and the large number (and average size) of subcontract assembly plants in the Philippines. While it is undoubtedly significant that by 1985, 61 locally owned plants employing a total of over 48,000 workers had emerged (not taking into account plants performing other specialized functions), we need to examine the technological bases of these plants, and the nature of their products, as well as, for instance, the extent to which they are likely to encourage the upgrading, in terms of skill, of their respective labour forces. These are among the elements I take to be essential for an assessment of the relationship between local production complexes, the emergence of 'autonomous' semiconductor industries, and the possibilities for genuine, lasting development.

We proceed by examining developments in particular territorial units and begin with the most advanced semiconductor production complex in Asia (other than Japan): South Korea.

South Korea

The emergence of electronics production in South Korea, began, rather like its Japanese counterpart, with an import substitution industry, protected by tariff barriers, in the early 1960s. Stimulated by foreign investment, particularly in semiconductors from the mid-1960s, indigenously owned electronics systems houses began to emerge. Beginning in the late 1970s, and stimulated by technology licensing arrangements with both American and Japanese companies, four semiconductor houses emerged as divisions of the Samsung, Goldstar, Hyundai, and Daewoo electronics and industrial conglomerates respectively.

All of these companies have opened R & D facilities in Silicon Valley and Hyundai, for a while, fabricated wafers there.[8] Their strategy, then, has been to acquire US technology and expertise in Silicon Valley, and then fabricate the wafers and assemble and test the semiconductors in Korea. Thus far, however, no Korean company has been able to produce the most technologically advanced ICs, and consequently they have concentrated on the memories market (principally 16K and 64K RAMs).[9] Indeed, one of the senior executives of Intel, which has a technology exchange agreement with Samsung, has suggested that the technology Samsung are being supplied with is 'at least two generations old' (quoted in Berney 1985: 51).

In addition to the electronics system houses which fabricate as well as assemble and test semiconductors, there are twelve subcontract assembly plants, including the largest in the world, Anam, with over 6,000 employees. As well as being the largest, Anam is also the most technologically sophisticated. As a result of some automation of its assembly lines, it is capable of bonding (assembling) complex, 148 pin ICs.[10] (Iscoff 1986).

Finally, the growth of the South Korean semiconductor complex can in a sense be gauged by the fact that the growing demand for wafers associated with it, has recently led to the indigenous production of silicon wafers under a joint-venture arrangement between the US producer, Monsanto and its Korean counterpart, Sanchok Industrial (UNCTC 1986: 397).

Whatever the combination of forces that have produced the emergence and success of the Korean semiconductor industry (as part of the industrialisation and spectacular economic growth of the country generally), it is undoubtedly the case that the national state has played a major role. It has done so in the following ways:

1. In addition to fostering the growth of the domestic consumer electronics industry by means of protectionist tariff barriers, in the early 1980s, the Korean government imposed an import ban on foreign-made electronics. This meant that the only way in which overseas electronic products could penetrate the growing, and increasingly lucrative domestic market, was by producing those goods inside Korea by means of wholly owned subsidiaries, joint-ventures, or licensing arrangements. The last two, in particular, have been utilized by foreign manufacturers, especially in semiconductor production (e.g. the Goldstar-AT&T joint-venture and the Intel-Samsung and INMOS-Hyundai licensing arrangements) (Berney 1985).
2. The Korean government has consistently provided substantial amounts of low interest capital for companies in 'targeted' sectors (and this includes electronics and semiconductors) rendering many companies dependent on government support. Consequently, while the Government has not acquired equity stakes in Korean corporations, it has ensured that industrial development has been jointly planned by the state and private capital in circumstances where the state is the dominant partner. (Berney 1985; Cummings 1987).
3. The Government heavily invests in electronics R & D. Under the auspices of the Korean Institute for Electronic Technology, the Government conducts basic semiconductor design work, which is then passed on to the various companies who adapt the designs for their own particular purposes. (Berney 1985; UNCTC 1986: 396).

The result of these co-ordinated initiatives by capital and the state is that according to Dataquest, the authoritative Californian/electronics market research firm, South Korea is the only one of Asia's newly industrialising countries (NICs) 'that has developed a manufacturing base to produce high value-added goods sufficient to sustain high growth rates over the next decade.' (quoted in Berney, 1985: 48).

Taiwan

The development of Taiwan's electronics and semiconductor industries parallels, in many ways, the emergence of their Korean counterparts. This is particularly the case with regard to state intervention. With wages beginning to rise by about 20 per cent annually (see Table 4.1) when productivity was rising 'only' by 8–11 per cent, the Government, beginning in the mid-1970s, began to encourage and assist technological upgrading of the electronics industry (Hong Kong Productivity Centre, *Electronics Bulletin* 2(1), 1982: 5–6).

State intervention took the form principally of the creation of the Electronics Research and Service Organisation (ERSO), one of the four divisions of Taiwan's Industrial Technology Research Institute (ITRI). ERSO, by virtue of a technology transfer arrangement with the US electronics systems house, RCA, became Taiwan's first local designer and fabricator of integrated circuits by the late 1970s (Neff 1980).

Taiwan effectively developed a nationalised semiconductor presence (in 1979), when ERSO encouraged the 'spin-off' of Taiwan's first commercial semiconductor company, the United Microelectronic Corporation, though state equity was subsequently reduced when United went public in 1985 (Goldstein 1985). In 1980 United was joined by Taiwan's largest electronics company, Tatung, when the latter began producing simple integrated circuits (Neff 1980).

More recently, six more semiconductor companies have emerged, including Sampo, Advanced Devices Technology, and the China Development Corporation. As with their Korean competitors, all of these latter companies, as well as United, Tatung, and ERSO, have invested in design facilities in Silicon Valley.

While the Taiwanese government continues to pump money into electronics R & D, for instance, planning to invest the equivalent of 3 per cent of electronics sales by 1989 (Hong Kong Productivity Centre, *Electronics Bulletin* 2(1), 1982: 5–6), the Taiwanese industry remains massively under-capitalised, compared to the Korean. Part of this is due to that fact that whereas Korean electronics (and semiconductor) production is organised in terms of large systems houses (Goldstar and Samsung) and industrial conglomerates (Hyundai and Daewoo), with the capital necessary for expensive R & D investments, the Taiwanese industry is not. In spite of the Government's attempts to encourage mergers, Taiwan's electronics firms still have an average size of only ninety-three employees (*Electronics Week,* 6 May 1985).

Singapore

Only Singapore, of the 'gang of four' has failed, so far, to develop an indigenous semiconductor production complex. While I am unable, at this point, to account fully for this phenomenon,[11] the Singapore government has been as interventionist in the electronics industry as have its Korean and Taiwanese counterparts. While it has not invested significantly in R & D, nor provided capital for indigenous 'start-ups', it has operated in various ways to encourage foreign firms to upgrade the technological basis of their operations. For instance, in 1979 the Government adopted a deliberate policy forcing-up labour costs. Rather than flee Singapore as a production base, foreign companies responded by upgrading their operations, but as a by-product, shedding unskilled labour, also. Given that Singapore was able to supply the quality engineering and technical labour essential for such technological upgrading, and at a relatively cheap cost, such Government intervention may well have been a major impetus for Singapore's emergence as a regional testing centre for US semiconductor firms. The technological upgrading of the Singaporean electronics industry can be seen in value-added data from the early 1980s. Whereas in 1978, 71.2 per cent of foreign direct investment went into capital-intensive, high value-added industrial processes, by 1981, the proportion had risen to 79 per cent (UNCTC 1986: 401).

More recently, the Singaporean government has attempted to turn the country into the software centre of Asia, having set up a number of training institutes with (partly) foreign capital and expertise. The best known and most developed of these is the Institute of Systems Analysis set up with assistance from IBM (Ernst 1983).

While no indigenous wafer fabrication plants have emerged in Singapore, one subcontract semiconductor assembler (TEAM Semiconductors) is now in operation using automated assembly lines (Iscoff 1986), as are one mask-making house (probably the only one in Asia outside of Japan), one producer of ceramic packaging material, and two firms which produce gold and aluminium bonding wires (UNCTC 1986: 405). In addition, local purchases of supplies by foreign semiconductor plants, seem to be increasing (Lim and Pang 1982).

Philippines

If there is evidence of the emergence of local semiconductor production complexes in most of the cores of the regional division of labour, what of the periphery? As I have already indicated, discernable developments in Malaysia and Thailand, are, in their best light, still in a stage of extreme infancy, while the possibilities of an Indonesian industry appear to have been stillborn. Among the peripheral units, only the Philippines would seem to hold out hope that a local production complex – along the lines of those in Korea, Taiwan, and Hong Kong – might emerge. However, while the Philippines has the largest concentration of subcontract assembly plants in the region (see Table 4.4) together with a number of small plants performing specialized tasks such as machining and 'burn-in' work – also on a subcontract basis (Scott 1987), it must be emphasized that there is no necessary connection between largely labour-intensive activities such as these, highly dependent as they are on the world-market fortunes of their customers, and the emergence of indigenous semiconductor plants with wafer fabrication and other more technologically advanced capabilities.

First, the majority of Philippine subcontract assemblers (including the largest amongst them) appear not to have automated their lines to any significant degree. As a result, they are only able to assemble technologically simpler semiconductors, such as 14–16pin ICs that can be done manually. Thus, for instance, a Hong Kong-owned Philippine assembler, Semiconductor Devices, assembles the simpler, 'low-end' product manually in the Philippines, but more sophisticated semiconductors, automatically, in Hong Kong (Iscoff 1986).

Second, as a major commentator on the industry has remarked:

> experience in assembly is of limited relevance to the manufacture of the heart of the device, the etched silicon chip. It is questionable just how much technology can be transferred without the educational and research infrastructure that is required for its successful application.... All the investment in the world will not transmit technology if the educated manpower required as a medium of transmission is lacking.
>
> (Grunwald and Flamm 1985: 116–17)

This absence of reliable supplies of engineering and technical labour would seem to be one of the primary reasons why the

Philippines and other peripheral units are unlikely to develop a more technologically advanced capacity in semiconductor production in the foreseeable future. Indeed, the absence of state intervention of the sort South Korea, Taiwan, and Singapore have engaged in, coupled with continuing internal political instability and world economic crisis, have resulted, if anything, in a serious crisis for the subcontract assemblers. In early 1986, the world's second largest subcontract assembler, Stanford Microsystems, collapsed and the next largest (in the Philippines), Dynetics, made three-quarters of its workforce redundant (*Global Electronics* 71, December 1986).

Malaysia

In 1970 the Malaysian government launched its New Economic Policy (NEP) which was designed to improve the economic status of the Malay majority in the population while eradicating poverty regardless of racial considerations. An important means to these ends was to be the industrialisation of the economy based, at least initially, on the attraction of foreign capital (Jomo 1986: 256–68). Consistent with practice elsewhere in the region (South Korea and Taiwan, for instance), export-processing zones were established, and by 1971, following a Government promotional drive in California, National Semiconductor became the first semiconductor company to establish an assembly plant in Malaysia. Subsequently the Government was able to attract investment by other US and more recently Japanese semiconductor houses, such that by early 1988, sixteen US firms had plants in Malaysia, helping to make it the world's third largest exporter of semiconductors after the US and Japan (Ehrlich 1988).

Given this concentration of production activity in Malaysia there are a number of additional points, both positive and cautionary, that need to be made. First, until recently, investment was overwhelmingly associated with the assembly of standardised memory ICs. Indeed, by 1986, Malaysia was assembling between 40 and 70 per cent of the world's share of the 'workhouse' ICs, the 64K RAM (random access memory) (Salih and Young 1987: 189). While investment in testing facilities and the automation of some assembly lines has begun to take place, until 1987, the 'Malaysian' semiconductor industry was in much the same situation as those of other peripheral economies in the region: it did not have the capacity to produce the technological cores of the semiconductor, nor engage in advanced R & D and design work. Most recently, however, as I have already mentioned, National Semiconductor has set up a small, though highly specialised wafer-fabrication facility in Penang, while other US (specifically Intel and Motorola) and Japanese (specifically Fujitsu) firms are contemplating investment in more generalised fabrication plants (Ehrlich 1988).

The emergence of wafer fabrication on the periphery of the regional division of labour (rather than, say, in Singapore or Hong Kong), is clearly a development of some significance. A factor which seems to have tipped the balance in Malaysia's favour, has been its recently generated capacity to supply sufficiently well-trained engineers and technicians at a cost significantly below that of even Singapore or Hong Kong, not to mention the United States etc. At National Semiconductor in Penang, for instance, 10 to 15 per cent of variable (i.e. non-capital) costs are made up of salaries for engineering and technical personnel, who currently (1988) receive an average, US$1,000 and US$600 per month respectively.[12]

While these tendencies towards capital deepening and the vertical integration of semiconductor production in Malaysia are undoubtedly encouraging from the point of view of Malaysia's continued economic development, there are a number of issues that remain problematic.

First, the semiconductor industry in Malaysia remains dependent both technically (in terms of design expertise and the testing of the more technologically advanced semiconductors) and managerially on Singapore and Hong Kong (depending on the company in question), and ultimately, of course, on corporate decision-making in the headquarter locations in the United States, Japan etc. This corporate dependency remains especially problematic in the light of the fact that Malaysia continues to concentrate on the production of standardised (rather than customised) memory devices which, in recent years, have been subject to notoriously unstable world market conditions (Ernst 1987; Salih and Young 1987).

Second, as with some of the other territorial units in the regional division of labour (such as Singapore), there is little evidence that a locally owned production complex has begun to develop. This situation persists in spite of the Malaysian government's intention (reflected in the NEP) to increase specifically Malay (rather than Malaysian Chinese or Indian) equity participation in the economy, including its industrial sector (Salih and Young 1987). Finally, the racialisation of the state and civil society in Malaysia, reflected not only in economic and public policy, but in the racial conflict of 1969 and the suppression of dissident opinion in 1987, must raise questions about political stability and therefore the internal climate for future investment (Jomo 1986; *Far Eastern Economic Review*, 14 April 1988).

While I shall deal in detail with the Hong Kong case in the next chapter, and with the likely evolution of the regional division of labour in Chapter seven, we must turn, at this point, to some more general comments on the role of the state intervention in industrialisation and economic growth in the region. This will be followed in the final section with comments on some of the social consequences of the growth of electronics and semiconductor production in East Asia.

The state and social change

In Chapter two, I suggested that the NIDL theorists have tended to devalue the extent to which a peripheral state theoretically had sufficient autonomy within the world system to affect, to some extent, the course and consequences of world-market oriented industrialisation. Their argument in fact seemed to be that the state did little more than provide for the infrastructural requirements of foreign manufacturers and the juridico-political context in which they could successfully 'super-exploit' their labour forces. While it is undoubtedly true that peripheral/semiperipheral states have made infrastructural provision and have developed repressive, and sometimes brutally enforced labour legislation (cf. Luther 1978; Deyo 1981 for Singapore; Sunoo 1978 for South Korea; and Villegas 1983 Chapter 5, for the Philippines under the Marcos regime), the situation is rather more complex than Fröbel and his colleagues (1980) and similiar theorists would have us believe.

First, I have already argued that the national states responsible for the cores of regional division of labour have intervened massively in order to assist the development of their respective electronics industries. When one recognises that these states (including Hong Kong) are precisely those that in the dream world of *laissez-faire* theorists are economic 'success stories' because of their supposed commitment to non-intervention, then one begins to get some measure of how ideological (not to mention empirically wrong) the dominant neoclassical paradigm in economics is. In spite of the fact that we are dealing with formally different states – repressive military dictatorships in South Korea (at least until 1987) and Taiwan, an authoritarian democracy in Singapore, and an autocratic colonial regime in Hong Kong – in all cases, economic development in these societies must now be seen, if anything, to be state-led (see Schiffer 1983 on Hong Kong; Lim 1983 on Singapore; Gold 1986 on Taiwan; Deyo 1987 on South Korea and Taiwan; Harris 1987 Chapters 2, 6 generally).

What is more, while those states have intervened in economic and social development in the interests of foreign and local capital, those same interventions have had positive consequences for the social and material well-being of workers. This is particularly the case in Hong Kong (Schiffer 1983; Castells 1986a; S. Y. Ho 1986) and Singapore (Lim 1983), both of which have developed extensive housing, educational, and welfare programmes. Indeed (and as I argue for the Hong Kong case in Chapter 5), their spectacular economic growth and rising living standards would have been inconceivable without such state intervention.

That said, the question of political order and stability in a peripheral state becomes increasingly important for foreign industrialists where they seek to technologically upgrade their operations. In situations where semiconductor firms invest, for instance, in labour processes that are dependent on large inputs of human labour (i.e. where they have low capital-labour ratios), then political stability is not as serious a problem as some theorists (e.g. Frank 1981) have assumed. When firms seek to increase capital–labour ratios, however, the political stability of the territorial unit takes on greater significance.

Repression, however, is not the only route to political stability, nor is it the most successful in the long run. 'Legitimation expenditure' on such things as public housing programmes coupled with representative democratic forms are often more reliable routes to political stability. In some of East Asia's 'hosts' to semiconductor and other foreign-owned industrial branches, however, their massive and growing material inequalities and foreign indebtedness coalesce to ensure that repression seems the only possible route to political stability in the context of a capitalist world system. The Philippines under Marcos was a case in point (cf. Bello *et al.* 1982; Villegas 1983) but it is far too early to say whether that country's recent democratisation is likely to bring with it greater stability that was previously the case.

Labour force composition

Evans and Timberlake (1980) have shown that 'dependent development' generates severely asymmetrical labour force compositions and rising relative inequalities in 'developing' countries. This situation tends to be particularly pronounced in urban areas where high proportions of the labour force are unskilled manual workers with significant numbers of them employed in the service sector. Smaller, though still significant numbers are employed in low-skilled white-collar jobs, with much smaller proportions in middle-class professional employment. Significantly in such urban labour markets, there tends to be little demand for skilled manual workers, and little emphasis on their training. With regard to those countries in which export-oriented industrialisation is a major part of their development strategy, very high proportions of their industrial labour forces are female. A recent United Nations Industrial Development Organisation (UNIDO) study has shown that in South Korea for instance, 75 per cent of all workers in export industries are female. In Malaysia's Beyan Lepas export-processing zone, 85 per cent of all production workers and nearly 100 per cent of all assemblers are women (UNIDO 1980). In the American-owned semiconductor industry, similar proportions of women in the manual labour force are evident not only in East Asian locations (Grossman 1979; Salih and Young 1987; Lin 1987; Chiang 1984) but also in the United States (Chapter 3; and Snow 1982), and as we shall see (in Chapter 6) in Scotland.

This preference for female labour in semiconductor and other world-market factories results, according to Sassen-Koob (1987), from the fact that women (particularly where they are rural-urban migrants) have already been culturally habituated to the acceptance of authority and low wages. Their employment in semiconductor plants, at least, has a number of important implications. First, it does appear to consitute an example of 'super-exploitation' of Third World workers by foreign enterprises. The women employed tend to be young; employment of those between sixteen to twenty-five years old tends to be the most usual (cf. Lim 1978b; Grossman 1979; Lin 1987) though my own research in Hong Kong suggests that child labour (girls as young as eleven or twelve) was probably utilised fairly extensively by US semiconductor firms during the 1960s. In addition, long working weeks continue to be the norm. In Hong Kong, for instance, as recently as 1983, an average working week of about 48 hours (including overtime) was still usual for workers in the electronics industry (Hong Kong Government 1983). Health risks also continue to pose a serious problem. Eye-sight and muscle-related ailments tend to be the most common, and usually result in the older women (i.e. those in their mid-twenties) being made redundant (Lim 1978b; Lin 1985).

The absorption of Third World women into semiconductor production has also had implications for traditional social relations. As Lim (1978b, 1982) and Lin (1987) have shown for Malaysia and Singapore for instance, traditional patriarchal social forms have been creatively utilized by many firms to assist the habituation of their female work forces to routine factory labour. Obversely, however, the entrance of women into wage labour has tended to disrupt traditional social relations. Its impact seems to have been double-edged. Whilst it appears to have weakened traditional patriarchal domination in the household, by providing the women with a degree of material independence, it has inevitably helped generate domestic conflict and on occasion placed female factory workers beyond the social and material support system of their families (Heyzer 1986). This situation has been particularly problematic where the women have been thrown out of factory jobs. There is now some evidence from a least one East Asian city of a link between expulsion from factory labour and involvement in prostitution (Phongpaichit 1981).

The final comment that is necessary at this point concerns the extent to which women semiconductor workers develop useful and transferable (i.e. 'saleable') skills from their employment. From her study of transnational electronics production in Malaysia and Singapore, Lim (1978a) has argued that employment in the industry has little beneficial effect on skill acquisition. Even in those sections of the East Asian industry where there has been some technological upgrading – such as with the introduction of automated testing equipment by Motorola in Hong Kong – work tasks remain largely unskilled. At Motorola, for instance, women are employed merely to manually feed integrated circuits into channels connected to computerised equipment where laser beams test for faults in the circuitry.

If the semiconductor industry – even in its most advanced East Asian locations – does little to upgrade the skills of those women who constitute, after all, the vast majority of its labour force, is it any better at encouraging the 'skilling' of its male workers? Other than the employment of small numbers of men in the most arduous unskilled tasks (such as in the warehousing and shipping sections), US semiconductor manufacturers employ men predominantly as supervisors, technicians, engineers, and managers. The numbers employed as technicians and engineers tends to be directly related to the capital intensity of the labour process. In Motorola's Hong Kong plant, for instance – one of the most capital-intensive American semiconductor facilities in East Asia – only about 300 of the 750 work force are manual production workers. The majority of the rest are technicians and engineers.

The availability of well-trained technicians and engineers in such places as Hong Kong and Singapore, (and most recently Malaysia) and at a much cheaper price than in the United States, has had a number of effects. First, it has led to the increasing employment of local engineers in the industry, to the extent that in the Hong Kong case local personnel are now in charge of at least three of the major American semiconductor operations in the territory. Second, as I have already argued, the availability of such labour is an important factor which bears on companies' decisions about whether to technologically upgrade their operation in a particular East Asian location.

And third, the semiconductor industry's (and more generally the electronics industry's) interest in employing local engineering graduates has encouraged state and private investment in educational resources designed to reproduce a labour force with these skills.

In summary, then, it seems that the relative dominance of a particular territorial unit within the international and regional divisions of labour in semiconductor production has few implications for their manual labour forces. Whether the women are employed by, say, National Semiconductor or Motorola in the United States or Scotland, Hong Kong or Malaysia, they are likely to remain the unskilled recipients of relatively low wages (while recognising significant differences in wage levels depending on the relative dominance of the territorial unit in the international and regional divisions of labour). For the employment of skilled (and predominantly male) technicians and engineers, however, the prospects seem to be more rosy. In this case, though, the possibilities for the skilling of the labour force and its continued reproduction, do depend on the relative dominance of the territorial unit within the international and regional divisions of labour. Where the unit has achieved core or semiperipheral status within the NIDL and where local electronics complexes have begun to develop, local educational institutes tend (often with state assistance) to adjust their curricula to the demands of the industry, and thus socialise the reproducion costs of a crucial sector of the semiconductor labour force (see Chapter 3). Once these processes are under way in particular locations within the regional division of labour, however, once local electronics complexes have begun to develop around semiconductor production, once local educational institutes begin to effectively produce the necessary technicians and engineers, the situation takes on a developmental logic of its own. The growth of such economic, social, and technological complexes in such parts of the regional division of labour as Hong Kong, South Korea, and Taiwan may present foreign semiconductor firms seeking to technologically upgrade their operations with very compelling reasons to upgrade their operations there, rather than in other locations. If this is indeed the case, then with the exception of Malaysia, we must be pessimistic about the possibilities for the upgrading of production in the territorial units currently on the periphery of the East Asian division of labour.

At this point in our study, we need to investigate in more detail some of the issues about the regional division of labour that this chapter has raised. We do this by examining how Hong Kong has emerged historically as one of the cores of the East Asian division of labour.

Chapter five

Hong Kong: The making of a regional core

In the previous chapter I argued that over the last seven or eight years or so, American semiconductor production in the developing societies of East Asia has been restructured such that a distinct, regional division of labour has begun to emerge. This regional division of labour, though ultimately subordinate to the overall international division of labour within the industry, has developed its own cores and peripheries with their respective sets of unequal transactions, dependencies, and developmental possibilities. In this chapter, we turn to examine in more detail the circumstances surrounding the emergence of Hong Kong as one of the cores of this regional division of labour. We begin the discussion by recounting the global and internal circumstances surrounding the emergence of Hong Kong as an industrialising society from the late 1940s. We then describe the development and key features of the territory's electronics and semiconductor industries. Subsequently, by reference to the articulation of world-system with internal elements, such as the role of the state and the nature of the labour force, we advance an explanation of the growth of these industries. Finally, we address the issue of the restructuring (including technological upgrading) semiconductor production has undergone in recent years, together with some of the social and developmental questions associated with it.

Hong Kong and the world-system

From its inception as a British colony in 1841, Hong Kong became increasingly incorporated into the world economy. Until the mid-twentieth century, however, its incorporation was largely based on its role as an East Asian entrepôt centre. Hong Kong, in other words, was a trans-shipment point for (largely) British exports to China (and to a lesser extent, other parts of the region) and for Chinese commodity and financial transactions with the core economies of Europe and the United States. In spite of the massive disruption of the entrepôt trade occasioned by the Japanese occupation of the territory, 1941–5, and the raging civil war on the Chinese mainland, as late as 1951, two-thirds of Hong Kong's exports (overwhelming re-exports at this point) were to China (Phelps Brown 1971: 2).

The manufacture of commodities had developed relatively early in Hong Kong. At least since the early 1930s, traditional carved wooden furniture had been produced for both domestic consumption and export (Cooper 1981), and basic cotton clothing was made for export to other parts of the British Asian empire, particularly Malaya (Mok 1969). Such manufactures, however, were produced in the traditional manner, using (in the former case, highly skilled artisan labour, and were never more than a small fraction of the territory's exports.[1]

The beginning of the industrial manufacture of commodities in Hong Kong from the early 1950s, was a result of a combination of three factors: the restructuring of the world economy after the Second World War, geo-political considerations associated with the Chinese Revolution of 1949 and the Korean War of 1950–53, and internal (to Hong Kong) social changes consequent to the Chinese Revolution. If, for the purposes of the current discussion, we can accept the over-accumulation/under-consumption theory of economic crisis (in its classic exposition by Baran and Sweezy 1966), then the Second World War, by absorbing surplus capital and physically eliminating surplus capacity, provided the basis for new rounds of accumulation subsequent to 1945. The post-war rejuvenation of the economies of the United States and (subsequent to the Marshall Plan) Western Europe, led to increasing real incomes in those core economies and hence consumer demands, at least from the early 1950s. While much of the rising demand for manufactured commodities was satisfied by production from within the core economies themselves, from about the mid-1950s onwards, this was not the case with the demand for cheaper cotton textiles and clothing in Britain. For a variety of structural reasons which we need not address here (but see Gregory 1985), the Lancashire cotton industry did not recover its pre-war dominance of the British market. There developed increasingly a shortfall between the demand for cotton products and the Lancashire industry's capacity to meet that demand at a cheap enough price. For reasons which we shall review later, part of that demand was met by Hong Kong's newly established textile industry.

The geo-political context in which Hong Kong was located by the end of the 1940s resulted in two inter-related economic and social consequences for the territory. They were, the virtual elimination of the entrepôt trade with China, and the migration of both labour and industrial capital and expertise from the Chinese mainland. The first of these developments was a direct consequence of the growing obsession of the US government and its allies with the supposed 'Communist threat' subsequent to the dissolution of the wartime alliance with the Soviet Union and the military triumph of the Chinese Communist Party over its Kuomintang opposition in 1949. With the outbreak of the Korean War in 1950 and Chinese support for its North Korea ally, in rapid succession the US government placed an embargo on all goods of Chinese origin and the United Nations prohibited the export of essential materials and strategic goods to China (Szczepanik 1958; Riedel 1972; So 1986). These political initiatives, together with the Chinese government's switch to the USSR as its principal trading partner (Lin, *et al.* 1980: 3), resulted in the virtual elimination of Hong Kong's entrepôt trade with the mainland which by 1952 had collapsed to barely one per cent of the territory's exports (Phelps Brown 1971: 2). In its wake came unemployment of the order of 15–34 per cent (Lin and Ho 1980: 5).

The collapse of the entrepôt trade would have resulted in serious economic crisis had it not been for the fact that since the late 1940s Hong Kong, as already mentioned, had begun to be the recipient of 'refugee capital and refugee labour that provided the impetus for [its] industrialisation' (So 1986: 244). Of particular significance was the transfer of industrial capital and managerial expertise in textile production from Shanghai. By the 1930s Shanghai had already developed a 'modern' cotton textile industry. By the late 1940s, with the declining military fortunes of the Kuomintang, many of its textile 'barons' had begun to transfer capital to Hong Kong, and to re-direct there new production machinery which otherwise would have been installed in their Shanghai plants (Mok 1969; Wong 1979). Within a relatively short time of arriving themselves in Hong Kong, the Shanghai textile barons, therefore, were able to set up modern factories combining the latest production equipment with cheap refugee labour. Though they themselves did not have the trading networks necessary for export purposes (for the Shanghai firms had supplied, predominantly, the Chinese domestic market), they benefited from the existence in Hong Kong of some 1,000 to 1,500 trading houses which previously had been involved in the entrepôt trade and as a consequence, had long-established links with the British and other export markets (Szczepanik 1958). By the mid-1950s, because of the problems associated with domestic cotton textile manufacture, British department stores and clothing chains began to seek out cheap foreign supplies of cloth and garments. Encouraged by Commonwealth preferential import tariff arrangements, they not only placed contracts with Hong Kong manufacturers, but directly assisted them to develop their production capacity and improve the quality of their output (Phelps Brown 1971: 12). From this point, it was a short step to foreign (again, initially British) direct investment inthe territory's textile industry, though it needs to be emphasized, that foreign investment has never been a major feature of this sector's development.

It was in these ways and for these reasons, then, that Hong Kong began to develop an industrial economy. While textiles (and garments) remain the principal industrial sector, in terms of employment and value of output through to the present day,[2] the latter part of the 1950s saw the beginnings of industrial diversification associated with the manufacture of plastic commodities (initially flowers and other decorations, but subsequently kitchenware also) and wigs, but, importantly for our purposes, electronic products. While the entrepôt trade revived once more in the 1950s, its supremacy in the Hong Kong economy was increasingly challenged by manufactured domestic exports. Throughout the decade, industrial productivity increased by an average of 20 per cent annually and by 1959 manufactured exports had surpassed the value of the entrepôt trade (Cheng 1985: 125). By the early 1960s Hong Kong had become the largest supplier of manufactured commodities in the developing world (Lin and Ho 1980: 2).

Growth of the electronics sector

Subsequent to the invention of the transistor in 1947 and the liberal licensing arrangements adopted by major US producers such as AT & T (Bell), General Electric and RCA, Japan was able to develop its own semiconductor industry from the early 1950s. Operating behind high tariff barriers, the bulk of the output fuelled Japan's own consumer electronics industry and led, in particular, to Sony's introduction of the transistor radio in 1955 (Okimoto, *et al.* 1984: 14). By the late 1950s, competition from US and domestic producers, and rising wages in Japan, resulted in Sony's partial internationalisation of radio assembly. They developed a joint-venture operation in South Korea (a Japanese colony until 1945) in 1958, and by 1959 had begun to assemble radios in Hong Kong under a subcontract arrangement. Soon Hong Kong's first electronics company, the Champagne Engineering Corporation, was assembling over 4,000 radios a month for Sony (Chen 1971). By 1960, Champagne and two other companies had begun to produce their own radios even more cheaply than the Japanese (because of lower labour and overheads costs). The result was that Hong Kong radio exports to the United States increased fifteen-fold during 1960–61, effectively undercutting their

Japanese competitors at the cheaper end of the market. By 1961, twelve firms were producing radios in the territory, of which two were joint-venture operations with US companies. The successful competition with Japanese producers resulted, in 1962, in a Japanese government ban on the export of transistors to Hong Kong. They were replaced by imports from Britain, but particularly from the United States (Chen 1971: Chapter 2).

The early development of Hong Kong as an offshore location for radio assembly by Japanese and US manufacturers, together with the rapid development of an indigenous production capability, provided the immediate context for investment by US semiconductor houses. As we saw in the previous chapter, Fairchild was the first semiconductor firm to develop production facilities anywhere in the Third World, which it did by setting up a transistor and diode assembly plant in Hong Kong in 1961. By 1966 when it moved to new premises in the territory's most important industrial area, Kwun Tong, Fairchild had become by far the largest electronics firm in Hong Kong, employing about 4,500 workers (Chen 1971). Since that time, Fairchild has been joined by nine other US semiconductor firms (which currently include National Semiconductor, Teledyne, Siliconix, Motorola, Sprague, Commodore, Microsemiconductor), two European (Philips and Ferranti) and two Japanese (Hitachi and Sanyo) producers.[3]

Before we examine the determinants of the development of 'Hong Kong's' semiconductor industry and the issues surrounding its recent emergence as a core of the regional division of labour, we need to convey some sense of the electronics industry's significance in the local economy and the extent of its dependence on foreign control, markets and the like.

Manufacturing in the Hong Kong economy

In order to gauge the economic and social significance of the electronics industry to the territory's development, we first need to indicate, in terms of some general indices, the contribution which manufacturing industry as a whole has made. Table 5.1 conveys a sense of the contribution manufacturing industry has made to Gross Domestic Product (GDP) relative to the contributions of other industries. Of particular note is the fact that manufacturing reached its zenith, in terms of its contribution to GDP, in the early 1970s (31 per cent in 1970), since when it has gradually declined to its present level of around 22 per cent. During the same period the contribution of the service sector expanded significantly. This expansion was partly due to the growth in tourism (which now constitutes the

Tables 5.1 Industrial structure of Hong Kong: Changing contribution to Gross Domestic Product (at factor cost), 1961–86 (percentages)

Industry	*1961*	*1970*	*1971*	*1972*	*1973*	*1974*	*1975*	*1976*	*1977*	*1978*	*1979*	*1980*	*1981*	*1982*	*1983*	*1984*	*1985*	*1986*
Agriculture and fishing[a]	3.7	2.7	1.9	1.7	1.6	1.6	1.5	1.4	1.4	1.3	1.1	1.0	0.9	0.9	0.8	0.6	0.7	0.6
Manu-facturing	23.6	30.9	28.1	26.9	26.6	25.8	26.9	28.3	27.2	26.9	27.6	23.9	22.8	20.6	21.9	24.1	21.9	21.9
Construction	6.2	4.2	4.9	5.3	5.5	6.1	5.7	5.4	6.0	7.0	7.3	6.7	7.5	7.3	6.0	5.3	5.0	4.5
Electricity, gas and water	2.4	2.0	1.9	1.7	1.6	1.8	1.8	1.7	1.4	1.5	1.1	1.3	1.4	1.9	2.0	2.5	2.7	3.0
Financial services	10.8	14.9	17.5	20.5	19.2	17.6	17.0	17.9	19.6	20.7	21.4	22.9	23.8	2.5	18.8	15.9	16.3	17.0
Other services	53.3	45.3	45.7	43.9	45.5	47.1	47.1	45.3	44.4	42.6	41.5	44.2	43.6	46.8	50.5	51.6	53.4	53.0
Totals	100.0	100.0	100.0	100.0	100.0	100.0	100.0	100.0	100.0	100.0	100.0	100.0	100.0	100.0	100.0	100.0	100.0	100.0

Sources: 1961: *Report on the National Income Survey of Hong Kong*, Hong Kong: Government Printer, 1969
1970–86: *Estimates of Gross Domestic Product*, Hong Kong Government
Data for 1961–83 adapted from Y. P. Ho (1986: 172)

Note: (a) plus very limited mining and quarrying

Table 5.2 Changes in the distribution of the Hong Kong labour force by industry, 1961–87 (percentages)

Industry	*1961*	*1971*	*1976*	*1981*	*1983*	*1984*	*1985*	*1986*	*1987*
Agriculture and fishing[a]	8.1	4.2	2.7	2.0	1.2	1.2	1.7	1.6	1.5
Manufacturing	43.0	47.0	44.8	41.2	36.6	37.8	35.2	34.5	33.3
Construction	4.9	5.4	5.6	7.9	8.3	7.9	7.7	7.8	8.1
Electricity, gas and water	1.1	0.6	0.5	0.7	0.6	0.5	0.6	0.7	0.7
Services	42.9	42.8	46.4	48.2	53.3	52.6	54.8	55.4	56.4
Totals	100.0	100.0	100.0	100.0	100.0	100.0	100.0	100.0	100.0

Sources: 1961: Report of the 1961 Census
1971: Hong Kong Publication and Housing Census: 1971 Main Report
1976: Hong Kong By-Census, 1976
1981–87: General Household Survey: Labour Force Characteristics
Data for 1961–84 adapted from Y. P. Ho (1986: 173)

Note: (a) plus very limited mining and quarrying

second most important contributor to GDP after manufacturing), and partly to the emergence of Hong Kong as a major international financial centre since the early 1970s.

Turning to the contribution manufacturing industry has made to employment creation, we can see from Table 5.2 that proportionately it was the leading provider of jobs during the 1960s and early 1970s (47 per cent in 1971, for instance) since when it has been progressively overtaken by the service sector, including government employment. In aggregate terms, manufacturing industry provided about 512,000 jobs in 1961, rising to a high of 990,000 by 1981 and falling by 1986 to about 918,000 (BBDO 1987: 3). In spite of its relative decline as an employer *vis-à-vis* other industries, Hong Kong still, in proportional terms, provides more jobs through manufacturing than any other country in capitalist East Asia.[4]

Hong Kong, like Singapore, never went through a phase of import-substitution industrialisation. Its manufacturing base, from the beginning, was almost entirely built on production for the world market. From the data collected in Table 5.3 we can see the changes in the relative contributions of the various manufacturing sectors to Hong Kong's export performance. It is immediately clear that Hong Kong's continuing development as an industrial society, at least in terms of its export performance, is narrowly based on two sectors, clothing and textiles (of which clothing is overwhelmingly dominant with 33.9 per cent of all manufactured exports as against 6.3 per cent for textiles in 1984; Y. P. Ho, 1986: 180) and electronics. Between them these two sectors now (1987) account for about two-thirds of the territory's domestic exports. While the contribution of the clothing and textile sector declined from 52 to about 44 per cent of exports been 1961 and 1987, the contribution of electronics has grown dramatically over the same period from 2.5 to nearly 23 per cent of exports (Table 5.3).

Though it is sufficient, at this point, merely to signal the narrow base on which Hong Kong's industrialisation rests, we will return to this point in Chapter seven, when we discuss the implications of the changing international division of labour for development prospects in the East Asian region. Here, however, we need to begin to shorten the focus of our analytic lens in order to investigate the electronics sector in more detail.

Electronics in general

We have already seen that the electronics sector has, in the last twenty years, established itself as the second most important contributor to domestic exports (Table 5.3). At this point we need to

Table 5.3 Export contribution of Hong Kong's manufacturing industries, 1961–87 (percentages by value)

Industry Sector	*1961*	*1963*	*1965*	*1967*	*1969*	*1971*	*1973*	*1975*	*1977*	*1979*	*1981*	*1983*	*1984*	*1985*	*1986*	*1987*
Clothing and textiles	52.1	53.0	51.9	48.6	47.1	49.9	50.4	54.0	47.4	43.3	41.8	39.7	40.1	43.0	43.2	43.7
Electronics	2.5	4.0	5.8	8.8	10.1	11.2	13.5	14.0	15.5	15.7	17.6	20.8	22.8	22.1	22.7	22.6
Watches, clocks and other precision instruments	0.7	0.8	0.8	1.3	1.7	2.0	2.5	3.9	6.6	9.4	10.3	9.2	7.8	8.8	9.1	8.7
Toys and dolls	4.6	5.5	7.5	8.3	8.5	8.6	7.7	6.5	8.0	7.5	8.1	7.7	7.7	7.6	7.6	6.5
Plastic products	6.8	6.6	5.9	4.9	4.0	3.4	3.5	1.8	2.1	1.9	1.6	1.7	1.8	1.7	2.0	2.1
Handbags and travel goods	0.7	0.8	0.9	1.2	1.4	1.7	2.1	2.0	2.0	2.2	2.0	1.7	1.5	1.2	1.0	0.9
Footwear	3.5	3.8	3.0	3.3	2.8	2.6	1.4	1.1	1.0	0.9	1.0	0.7	0.8	0.8	0.8	0.7
Others	25.1	21.9	21.1	20.7	21.7	18.2	16.3	14.0	14.5	16.5	15.1	16.2	15.3	14.8	13.6	14.8
Totals	100.0	100.0	100.0	100.0	100.0	100.0	100.0	100.0	100.0	100.0	100.0	100.0	100.0	100.0	100.0	100.0

Sources: Hong Kong Trade Statistics (various issues) and
Hong Kong Review of Overseas Trade (various issues)
Data for 1961–84 adapted from Y. P. Ho (1986: 180)

examine its contribution to employment creation and the broad outlines of its control structures and markets before we deal directly with the semiconductor branch.

Table 5.4 Establishments and employment in the Hong Kong electronics industry, 1961–87

	Establishments		*Number of employees*			
Year	Number	Average size (by employment)	M	F	F%	Total
1961	3	36.0	31	77	(71.3)	108
1962	14	62.9	262	619	(70.3)	881
1963	18	67.7	308	910	(74.7)	1,218
1964	29	101.4	621	2,320	(78.9)	2,941
1965	29	149.9	1,142	3,205	(73.7)	4,347
1966	39	278.8	1,897	8,976	(82.6)	10,873
1967	63	313.5	3,338	16,410	(83.1)	19,748
1968	100	231.3	4,308	18,825	(81.4)	23,133
1969	114	274.2	6,573	24,686	(79.0)	31,259
1970	173	211.3	8,500	28,052	(76.8)	36,552
1972	280	146.6	10,525	30,531	(74.0)	41,056
1974	382	143.7	12,881	42,002	(76.5)	54,883
1975	387	108.5	10,731	31,240	(74.4)	41,979
1976	680	88.4	12,948	47,161	(78.5)	60,109
1977	711	98.7	19,057	51,131	(72.8)	70,188
1978	793	93.0	21,439	52,297	(70.9)	73,736
1979	1,041	74.6	19,294	58,369	(75.2)	77,663
1980	1,104	81.4	28,066	61,773	(68.8)	89,839
1981	1,216	73.1	28,609	60,312	(67.8)	88,921
1982	1,305	65.9	27,605	58,341	(67.9)	85,946
1983	1,351	70.1	31,430	63,201	(66.8)	94,631
1984	1,441	73.9	35,866	70,547	(66.3)	106,413
1985	1,284	63.9	27,712	54,283	(66.2)	81,995
1986	1,090	69.0	26,334	48,844	(65.0)	75,178
1987	1,180	66.6	27,892	50,670	(64.5)	78,562

Sources: 1961–70: Hong Kong Labour Department Annual Reports
1972–81: Electronics Industry Data Sheets, Hong Kong Productivity Centre (various issues)
1982–87: Report of Employment and Vacancies Statistics, Census and Statistics Department, Hong Kong Government

Table 5.4 provides data on the growth of the territory's electronics industry since the early 1960s. Within twenty years of its emergence as a subcontract assembler of cheap transistor radios, the sector had grown to encompass the production of a vast range of consumer items (TVs, VCRs, tape-recorders, hi-fi equipment, computers etc.) as well as components such as semiconductors, capacitors, condensers, transformers etc. By the late 1980s the industry was spread across nearly 1,200 factories employing over 78,000 people.[5]

There are, perhaps, two things of particular interest about the data in Table 5.4. First, the table shows that by the late 1960s, the industry was organised in terms of medium-sized factories of 200 to 300 employees. The rapid proliferation of factories during the 1970s and 1980s, however, coincided with a significant reduction in average size. This development has undoubtedly been associated with a proliferation of the subcontract assembly of particular parts of the products to small (often no more than 10 workers), family-run firms (Sit, *et al.* 1979), as well as a major growth in undocumented informal sector subcontracting (Sit 1983). Second, and an issue to be addressed later in this chapter, Table 5.4 indicates that the electronics workforce in Hong Kong is overwhelmingly female, ranging from a high of 83 per cent of the total in the late 1960s to about 65 per cent at the present time. The gender composition of the electronics workforce in Hong Kong, then, is very similar to that of the industry in California (Chapter 3), East Asia generally (Chapter 4) and as we shall see, in Scotland also (Chapter 6).

We now turn to examine the extent of foreign control over the territory's electronics industry. From the data collected in Table 5.5, it is clear that throughout the period 1971–87, electronics has been the recipient of more foreign direct investment than any other manufacturing sector. Indeed, by 1987 electronics was receiving nearly four times as much direct investment than the next most significant sector, construction. In spite of this relative growth in foreign direct investment in electronics, it has not resulted in significant job creation. On the contrary, while aggregate employment resulting from foreign investment has remained fairly stable, or indeed increased, proportionately it has declined relative to other sectors. Thus, while in 1971 foreign investment in electronics created 52.7 per cent of all manufacturing employment created by foreign direct investment, by 1976 this had dropped to 39.0 per cent, by 1983 to 30.1 per cent, though by 1987 it had increased again to 37.3 per cent (derived from Table 5.5). If for the sake of comparison, we look at the territory's principal manufacturing sector (in terms of employment and value of exports), and until 1985 the second most important recipient of foreign direct investment, clothing and textiles, we find a very different picture. While proportionately foreign investment in textiles has been reduced by two-thirds between 1973 and 1987, aggregate employment created by that investment has more

Table 5.5a Foreign direct investment in Hong Kong's manufacturing industry, 1971–8

Industry sector	1971			1972			1973			1974			1975			1976			1977			1978		
	A	B	C	A	B	C	A	B	C	A	B	C	A	B	C	A	B	C	A	B	C	A	B	C
		%			%			%			%			%			%			%			%	
Electronics	61	34.9	29,773	60	31.7	29,122	51	21.8	31,538	56	35.6	34,656	61	34.7	24,146	65	30.7	24,552	61	26.1	29,620	68	24.8	29,803
Clothing and textiles	65	21.2	10,942	68	20.8	12,673	51	22.0	10,676	61	13.7	12,054	71	15.0	12,372	82	13.7	13,539	87	15.8	17,176	95	15.0	17,748
Chemicals	5	1.5	80	5	1.3	154	8	8.9	608	11	5.8	633	11	5.7	639	12	13.1	739	13	11.9	587	19	12.2	817
Electrical products	6	2.1	1,145	10	2.3	1,337	8	3.6	1,719	11	5.7	1,769	14	5.7	3,847	18	5.0	4,203	21	10.2	4,918	25	9.7	5,217
Printing and publishing	5	3.0	763	5	3.7	910	6	3.1	1,151	8	3.7	1,207	8	3.6	1,182	8	3.1	1,182	9	7.1	1,777	10	6.7	1,857
Watches and clocks	12	3.4	3,186	13	3.2	4,032	12	13.2	4,649	17	10.9	4,643	20	11.1	5,391	23	9.8	5,577	25	6.8	4,894	27	6.5	5,024
Metal products	10	2.6	1,852	9	2.2	725	9	1.8	928	12	3.0	1,071	12	2.9	1,434	17	2.9	1,815	26	4.8	2,215	30	4.7	2,215
Toys	12	3.8	7,034	9	3.6	9,241	8	5.3	5,856	7	3.5	5,857	8	3.4	4,412	8	3.0	4,412	10	3.1	5,392	11	3.2	5,742
Food	6	0.7	215	7	6.8	1,115	5	4.2	1,184	6	3.6	1,206	6	3.5	1,533	8	3.9	1,657	14	3.7	1,632	17	5.8	1,967
Construction	NA	NA	NA	5	2.2	226	6	2.3	545	4	3.1	545	4	3.1	333	5	2.8	453	5	1.7	613	5	1.6	613
Others	60	25.8	1,529	58	22.2	1,735	51	13.8	1,342	54	11.2	1,697	55	11.3	4,318	65	12.0	4,871	65	8.8	5,943	79	9.8	7,153
Totals	242	100.0	56,519	249	100.0	61,270	221	100.0	61,216	247	100.0	66,338	271	100.0	59,602	311	100.0	63,000	339	100.0	74,758	386	100.0	78,330

Sources: 1971–75: Hong Kong Department of Industry
1976–78: Hong Kong Department of Trade, Industry and Customs, Annual Statistical Reviews

Notes: A: Number of establishments
B: Proportion of total foreign manufacturing investment by sector (%)
C: Number of employees

Table 5.5b Foreign direct investment in Hong Kong's manufacturing industry, 1979–87

Industry sector	1979			1980			1981			1982			1983			1984			1985			1986			1987		
	A	B	C	A	B	C	A	B	C	A	B	C	A	B	C	A	B	C	A	B	C	A	B	C	A	B	C
		%			%			%			%			%			%			%			%			%	
Electronics	72	23.0	30,891	75	21.6	28,889	57	38.0	27,676	64	36.5	29,114	78	35.9	29,630	82	35.5	34,824	99	36.1	33,731	115	32.1	32,254	124	42.0	37,853
Clothing and textiles	100	15.7	18,553	105	16.3	22,089	87	12.4	19,242	95	11.6	20,234	103	11.5	21,946	89	9.7	17,205	99	10.2	19,737	105	8.6	16,625	113	7.5	22,185
Chemicals	22	13.2	1,144	23	12.0	1,167	21	7.2	1,312	23	10.5	1,445	24	6.6	1,381	23	7.0	1,625	26	6.8	1,676	27	4.9	2,004	28	6.3	2,166
Electrical products	29	9.0	5,587	34	10.6	6,094	29	7.5	8,878	32	6.3	8,333	34	6.1	8,546	38	8.0	9,569	39	7.7	9,619	32	6.2	6,991	35	5.3	8,550
Printing and publishing	12	6.7	1,982	12	6.2	2,029	11	2.6	1,538	11	2.4	1,532	12	2.4	1,560	15	5.8	2,018	15	5.7	2,092	18	7.3	2,202	20	3.2	3,451
Watches and clocks	33	7.3	6,267	41	7.1	7,228	35	5.7	8,864	41	5.1	9,125	47	5.0	8,778	30	4.8	4,086	35	5.1	4,352	32	5.1	4,142	29	5.7	4,193
Metal products	32	4.5	2,385	34	4.1	2,478	30	3.6	2,337	30	3.3	2,337	33	3.6	2,900	44	4.8	3,651	47	4.7	3,724	42	4.5	5,361	37	3.8	5,911
Toys	12	2.9	6,058	11	2.6	5,853	13	3.3	6,337	13	1.8	4,738	13	1.6	4,738	11	0.9	3,090	12	1.0	3,190	12	1.1	3,347	18	1.2	3,302
Food	17	5.3	2,376	18	4.8	2,384	19	5.5	3,169	19	5.1	3,169	23	5.9	3,667	19	7.2	3,559	21	7.0	3.607	20	5.2	3,658	19	4.1	3,640
Construction	5	1.5	613	5	1.4	613	10	2.7	985	12	10.5	1,445	13	10.1	1,495	9	7.6	1,429	11	7.4	1,503	8	13.6	1,059	8	10.9	1,054
Others	93	10.9	7,526	102	13.4	8,476	83	11.5	9,801	97	11.1	11,476	105	11.3	11,495	82	8.7	7,977	91	8.3	8,349	98	11.4	8,783	110	10.0	9,420
Totals	427	100.0	83,382	460	100.0	87,282	395	100.0	90,229	438	100.0	92,803	486	100.0	96,046	442	100.0	89,033	495	100.0	91,580	509	100.0	86,426	541	100.0	101,455

Sources: 1979–83: Hong Kong Department of Trade, Industry and Customs, Annual Statistical Reviews
1984–87: Report on the Survey of Overseas Investment in Hong Kong Manufacturing Industries

Notes: A: Number of establishments
B: Proportion of total foreign manufacturing investment by sector (per cent)
C: Number of employees

than doubled over the same period, though in proportion to total manufacturing employment created by foreign investment, it has remained relatively stable, fluctuating only between 19.4 and 21.9 per cent over the 1971–83 period (derived from Table 5.5). While we will return to the significance of these changes in the electronics sector later in the chapter, it may be useful to mark, at this point, the fact that the data in Table 5.5 may suggest that capital deepening has occurred in the foreign-owned electronics companies in Hong Kong, at least *vis-à-vis* their counterparts in the textile industry.

While it is clear that the bulk of foreign direct investment in manufacturing has been increasingly directed towards the electronics sector, to what extent is the electronics industry as a whole controlled by foreign interests? By comparing the data presented in Tables 5.4 and 5.5, we can develop an estimate of the situation. Table 5.6 displays this estimate. The data collected there show that in terms of factory ownership and the proportion of the workforce which those factories employ, foreign control of the Hong Kong electronics industry declined significantly during the decade and a half from the early 1970s to the late 1980s. This evidence of decline in foreign control must be tempered, however, with the recognition that in terms of investment, foreign control has probably remained fairly stable.[6] What is equally significant about the data in Table 5.6, however, is that as we noted for South Korea and Taiwan in Chapter four, it demonstrates the substantial growth, in Hong Kong, of a locally owned electronics complex. The growth of the locally owned complex is emphasized, additionally, by the persistent decline in the average size of all factories relative to the much larger and relatively stable average size of the foreign-owned factories (Table 5.6; see Sit *et al.* 1979, on the very small size of the majority of Hong Kong-owned electronics factories). The development of a locally owned electronics production complex in Hong Kong has been significant for the territory's emergence as a core of the East Asian regional division of labour in semiconductor production. Consequently, I shall return to this issue later in the chapter.

While I have conveyed the broad parameters of foreign control of the Hong Kong electronics industry, it is necessary at this point, and, of course, crucial to the subject matter of the book to explore the national origin of that control. Relevant data for this purpose are assembled in Table 5.7.

If we can gauge foreign control in terms of indices such as the number of factories owned, the proportion of investment, and the proportion of the workforce employed, then the United States and Japan dominate the foreign-owned sector of manufacturing generally, and electronics in particular. Between them, they supplied

Table 5.6 Foreign control of the electronics industry in Hong Kong: Selected years, 1972–87

Year	Number of establishments	Proportion foreign owned	Total workforce	Proportion working in foreign-owned establishments	Average size of all establishments	Average size of foreign-owned establishments
		%		%		
1972	280	21.4	41,056	72.4	146.6	488.1
1974	382	14.7	54,883	63.1	143.7	618.9
1975	387	15.8	41,979	57.5	108.5	395.8
1977	711	8.6	70,188	42.2	98.7	485.6
1978	793	8.6	73,736	40.4	93.0	438.3
1980	1,104	6.8	89,839	32.2	81.4	385.2
1981	1,216	4.7	88,921	31.1	73.1	485.5
1982	1,305	4.9	85,946	33.9	65.9	454.9
1983	1,351	5.8	94,631	31.3	70.1	379.9
1984	1,441	5.7	106,413	32.7	73.9	424.7
1985	1,284	7.7	81,995	41.1	63.9	340.7
1986	1,090	11.0	75,178	42.9	69.0	280.5
1987	1,180	10.5	78,562	48.2	66.6	305.3

Sources: As for Tables 5.4 and 5.5.

Table 5.7 Foreign investment in the electronics and manufacturing industries of Hong Kong: Country sources, 1981, 1984 and 1987

	1981						1984						1987					
	A		B		C		A		B		C		A		B		C	
Country Source	All manu-facturing	Elec-tronics	All manu-facturing	Elec-tronics	All manu-facturing	Elec-tronics	All manu-facturing	Elec-tronics	All manu-facturing	Elec-tronics	All manu-facturing	Elec-tronics	All manu-facturing	Elec-tronics	All manu-facturing	Elec-tronics	All manu-facturing	Elec-tronics
			%	%					%	%					%	%		
USA	111	30	43.7	63.6	41,227	19,176	124	43	53.7	83.3	39,708	21,397	163	58	44.0	73.7	38,276	18,661
Japan	111	43	31.6	33.6	23,066	3,793	117	14	21.0	10.3	18,574	3,891	134	26	19.0	10.4	22,635	5,559
UK	43	4	6.9	1.4	9,909	2,232	52	9	6.9	5.7	13,407	4,595	55	8	5.9	2.1	13,401	4,981
Netherlands	7	2	2.7	N/A	2,872	NA	10	2	2.8	N/A	4,462	N/A	11	7	2.8	5.3	4,732	3,797
Switzerland	20	-	2.8	-	3,314	-	19	2	2.5	N/A	3,547	N/A	23	1	1.7	-	1,832	-
Denmark	4	-	2.3	-	564	-	2	-	N/A	-	N/A	-	4	-	0.7	-	644	-
Australia	21	-	2.2	-	2,672	-	25	1	1.6	N/A	2,927	N/A	18	2	1.2	-	3,188	-

Singapore	15	2	1.8	N/A	1,818	N/A	18	N/A	2.1	N/A	2,286	N/A	22	1	1.2	–	3,332	–
Taiwan	23	2	1.4	N/A	2,775	N/A	17	N/A	0.9	N/A	1,416	N/A	12	1	0.4	–	757	–
France	5	–	1.1	–	595	–	5	–	0.7	–	541	–	8	–	0.5	–	842	–
Philippines	4	–	0.7	–	1,230	–	7	–	2.6	–	1,146	–	7	2	1.7	–	1,310	–
West Germany	21	–	0.5	–	2,120	–	33	3	1.0	0.1	3,285	243	32	5	1.1	0.4	4,075	564
Thailand	11	–	0.5	–	780	–	12	–	0.4	–	772	–	15	–	0.6	–	1,439	–
Others	42	5	2.0	N/A	4,554	N/A	61	21	2.2	0.6	11,270	5,328	114	22	19.2	8.1	18,917	6,861
Totals	438 (391)[a]	62	100.2	100.0	97,496 (90,059)[a]	25,201[b]	502	95	98.4	100.0	103,341 (89,033)[a]	35,454 (34,824)[a]	618 (541)	133 (124)	100.0	100.0	115,380 (101,455)	40,463 (37,853)

Source: Reports on the Surveys of Overseas Investment in Hong Kong's Manufacturing Industry, 1981, 1984 and 1987. Hong Kong Department of Trade, Industry and Customs

Notes: A: Number of establishments
B: Proportion of total foreign investment
C: Number of employees

(a) Figures in parentheses denote the actual number.
Discrepancy reflects the fact that double counting occurred when more than one overseas source invested in the same company.
(b) Underestimate due to unavailable data.

around 75 per cent of foreign direct investment in the manufacturing sector in 1981 and 1984, though this had fallen to 63 per cent by 1987. Additionally they provided 66, 56, and 60 per cent of jobs in the foreign-owned sector in those years respectively (derived from Table 5.7). The presence of US and Japanese business interests in the electronics industry has been even more pronounced. In 1981, they provided 97.2 per cent of foreign direct investment and 91.1 of foreign sector jobs in the industry. The equivalent proportions for 1984 were 93.6 and 71.3 per cent and for 1987, 94.1 and 64.0 per cent respectively. Also of significant interest is the fact that in 1987, while the US and Japan owned only 6 per cent of Hong Kong's electronics factories, together they employed nearly 31 per cent of the industry's total workforce. In fact, US companies alone employed nearly 24 per cent of the workforce (derived from Tables 5.4 and 5.7).

Comparing the data for 1981, 1984, and 1987 (in Table 5.7) it is clear that while the Japanese presence relative to other foreign interests has declined in manufacturing generally and in electronics in particular (in terms of investment, though not the workforce, at least in electronics),[7] the US presence generally has increased, or at least remained stable. For the manufacturing sector as a whole US investment, for instance, increased from about 44 to 54 per cent of all foreign direct investment between 1981 and 1984 though by 1987 had dropped again to 44 per cent. In the electronics sector US investment increased from about 64 to 83 per cent of the foreign direct totals between 1981 and 1984 and by 1987 amounted to about 74 per cent. Of particular note for our later discussion, however, is the fact that although US investment in the electronics industry more than tripled between 1981 and 1987 (from US$207.3 million to US$707.1 million at current prices),[8] it did not result in an increase in employment (see Table 5.7). Although these data should be treated with caution, they would seem to imply that US electronics firms in Hong Kong may be involved in a process of capital deepening, and hence, presumably, an upgrading of the technological sophistication of their labour processes. As with our earlier comment in this vein, we shall return to this issue in a subsequent section.

Before we move on, finally, to investigate the semiconductors branch of the Hong Kong electronics sector, we need to indicate the destination of electronics exports from the territory. Table 5.8 provides data on the share of exports consumed by Hong Kong's principal electronics markets. While the United States was, and remains, the dominant market for Hong Kong's electronics products with an average annual share of 44.4 per cent (1976–87), the principal EEC markets (UK, West Germany, France, and the Netherlands) are also particularly important (other than the 'rest of the world') with an

Table 5.8 Market share of electronics exports from Hong Kong, 1976–87 (percentages by value)

Market	*1976*	*1977*	*1978*	*1979*	*1980*	*1981*	*1982*	*1983*	*1984*	*1985*	*1986*	*1987*
USA	49.9	48.4	43.7	36.0	38.1	44.4	46.3	50.5	47.4	45.5	44.0	38.2
West Germany	14.5	12.8	13.9	14.2	11.7	8.6	7.2	6.8	5.8	5.3	6.9	6.9
UK	4.5	4.4	9.0	10.2	8.2	8.4	6.1	5.7	7.2	5.2	4.7	5.4
China	–	–	–	–	2.6	4.1	4.7	4.0	9.8	15.1	9.8	12.3
Netherlands	3.0	3.0	3.9	4.0	3.3	3.2	3.1	3.0	2.6	2.6	2.9	3.8
France	0.9	1.4	2.0	2.6	3.0	2.7	2.3	2.0	1.5	–	2.2	3.4
Singapore	2.0	2.1	2.4	3.0	3.1	1.5	1.6	1.7	2.1	2.1	2.6	3.3
Japan	1.6	1.3	1.4	2.2	1.7	1.5	1.7	2.2	2.3	2.1	2.7	2.8
Rest of the World	23.6	26.4	23.7	27.9	28.3	25.6	27.0	24.1	16.1	22.1	24.2	23.9
Totals	100.0	100.0	100.0	100.0	100.0	100.0	100.0	100.0	100.0	100.0	100.0	100.0

Sources: Trade Statistics, Hong Kong Department of Census and Statistics, various years. Adapted from Hong Kong Productivity Centre (1984, Table 1.5: 92) 1982–7: *Hong Kong Trade Review*, Hong Kong Trade Development Council

annual average of 21.3 per cent over the 1976–87 period. Rather like the structure of the Hong Kong manufacturing sector generally, and the concentration of foreign control over its electronics industry, the territory's markets for its electronics products are very narrowly based, with an annual average of nearly two-thirds of its exports being consumed by the US and the EEC.[9]

There are two things of subsidiary interest which appear from the data in Table 5.8. The first is the very small export market for Hong Kong's electronics products constituted by Japan (an annual average of only 2 per cent, 1976–87). These data would suggest that not only do local electronics manufacturers have difficulty penetrating the Japanese market, but that Japanese companies who manufacture in the territory, produce mainly for the world, not the Japanese market. Second, we should mark the emergence of China as the second largest national export market since the mid-1980s. While this development is consistent with China's re-emergence as, in effect, Hong Kong's domestic market, it should be borne in mind that it is unlikely that China can be anything other than a market for technologically low-grade electronic products for the forseeable future. We shall develop these last comments again in Chapter seven. At this point, however, we turn directly to the development of the Hong Kong semiconductor industry.

Semiconductors in particular

Fairchild Semiconductor, the 'mother' of Silicon Valley, was the first semiconductor firm in the world to seek to reduce its production costs by means of internationalising the most labour-intensive part of its production process, namely assembly. As I have mentioned already, Hong Kong became its first international location in 1961. While I shall deal with the determinants of semiconductor production in Hong Kong in the next section, it is necessary here to bear in mind that, as I indicated in Chapter four, Fairchild moved offshore in the context of substantial Japanese price competition in transistor markets. Japanese manufacturers were able to undercut their US competitors because of access to cheap domestic labour. The response of US manufacturers took one of two forms. Like Fairchild, they either moved assembly offshore in search of unskilled labour even cheaper than that to which Japanese companies had access, or they opted for automated assembly in the United States. At that time (late 1950s-early 1960s) semiconductor technology was subject to rapid cycles of innovation and hence short production runs, and re-programmable assembly technology was not yet available. As a result, automated assembly proved uneconomic and firms

Table 5.9 Principal sources of assembled semiconductor devices imported into the United States under Tariff Items 806.30 and 807.00

Year	*Hong Kong* %	*Indonesia* %	*Korea* %	*Malaysia* %	*Philippines* %	*Singapore* %	*Taiwan* %	*Thailand* %	*All Asia* ($000,000)[1]
1969	49.2	–	22.9	–	–	9.8	14.8	–	228
1970	44.6	–	23.2	–	–	17.9	8.9	–	254
1971	32.7	–	30.9	–	–	23.6	12.7	–	270
1972	25.4	–	26.9	–	–	37.3	10.4	–	452
1973	19.4	–	23.6	8.3	1.4	33.3	12.5	–	740
1974	17.1	–	22.9	21.4	2.9	22.9	12.9	–	977
1975	11.8	–	17.1	30.3	5.3	26.3	7.9	–	858
1976	13.1	–	20.7	25.6	7.3	28.0	7.3	–	1,240
1977	8.0	1.1	21.8	27.6	6.9	24.1	9.2	1.1	1,567
1978	6.8	1.1	17.0	34.1	9.1	22.7	5.7	3.4	1,948
1979	4.6	2.3	13.8	33.3	11.5	23.0	4.6	2.3	2,212
1980	4.5	2.3	10.2	34.1	15.9	25.0	4.5	3.4	2,518
1981	3.4	2.3	9.2	34.5	18.4	23.0	4.6	4.6	2,536
1982	3.4	2.2	8.9	36.0	20.2	19.1	4.5	3.4	2,800
1983	1.2	2.4	16.5	36.5	21.2	12.9	4.7	4.7	2,876

Source: Calculated by Scott (1987, Table 4: 146) from Table 3–7 in Grunwald and Flamm (1985).
Note: 1. All values in constant 1983 US $

who opted for that 'solution', such as Philco, suffered disastrous consequences (Grunwald and Flamm 1985: 69–70). The 'Fairchild solution', however, proved highly successful, and as a result, as we saw in the previous chapter, US merchant producers in rapid succession began to invest in assembly facilities in various parts of East Asia. At the high point of the US semiconductor presence in Hong Kong (early to mid-1970s), eight companies had assembly plants which employed probably in the region of 15,000 workers.[10] Fairchild alone at that time employed in excess of 4,500 people (UNCTC 1986: 383–9).

By the late 1960s, Hong Kong had become the principal Asian assembler of semiconductors for the US market. By 1972 however, it had begun to be overtaken by South Korea and Singapore. By 1983 its exports to the US market had slipped to only 1.2 per cent of the total for the developing countries of East Asia, even falling behind Indonesia, which, with only two plants (in 1983) has the least developed semiconductor industry in the region (see Table 5.9).

This relative decline in exports to the United States, however, is not evidence of a decline in the industry itself. On the contrary, the data collected in Table 5.10 would seem to imply the opposite. What this may be circumstantial evidence of, however, is an alteration in the type of labour processes that have emerged in the territory since the late 1970s, such that proportionately less 'assembly' work is done there relative to other parts of the region. Let us explore this hypothesis further.

In the previous chapter, I developed a general argument that in recent years US companies operating in the developing countries of East Asia had been diverting larger proportions of their investment in assembly processes to the periphery of the regional division of labour, while investment in more capital-intensive, technologically advanced labour processes (such as final testing) had tended to locate in the gang of four, but in Singapore and Hong Kong in particular. While the previous chapter explored some of the general reasons for this phenomenon, and we shall examine these again, specifically in relation to Hong Kong in the next section, here we need to convey some sense of the nature of semiconductor manufacturing operations which currently exist in the territory.

The first thing that needs to be emphasized is that Hong Kong is no longer a production site which specializes in the *assembly* of semiconductors. As the data collected in Table 5.10 shows, the re-export of integrated circuits and transistors/diodes which are imported into Hong Kong already assembled, now exceeds substantially the export of devices assembled in the territory. While a proportion of re-exports may result from the fact that Hong Kong is the

Table 5.10 Hong Kong: Exports of semiconductors, 1974–87 (thousands of US $)

	Domestic exports		*Re-exports*	
Year	ICs	Transistors and diodes	Mounted ICs	Transistors and diodes
1974	46,640	78,819	2,230	14,257
1975	32,373	51,440	2,582	13,307
1976	56,923	75,112	5,564	29,745
1977	54,388	82,774	11,235	30,119
1978	63,086	89,736	41,432	32,688
1979	89,329	87,106	82,458	59,216
1980	101,754	125,341	110,136	94,932
1981	113,840	98,635	132,359	189,084
1982	89,098	90,415	140,810	102,932
1983	82,771	62,698	273,615	92,534
1984	132,131	67,792	487,209	116,142
1985	106,918	71,493	403,414	103,581
1986	82,276	77,919	498,398	121,620
1987	99,513	103,360	811,105	168,633

Sources: Hong Kong Review of Overseas Trade, (various editions).
Data for 1974–81 adapted from UNCTC (1986, Table VIII, 1: 385)

regional marketing headquarters of a number of US semiconductor firms (specifically Motorola, Sprague, and Zilog), and hence could be used merely as a trans-shipment point, it is highly likely that the data in part reflects the fact that the territory has emerged as a centre for the final testing of a variety of ICs and transistors assembled elsewhere in the region.

Table 5.11 indicates the type of labour processs currently operated by Hong Kong's semiconductor manufacturers. If we compare the situation in Hong Kong with the labour processes in other parts of the East Asian regional division of labour evident in Figure 4.2, it is clear that Hong Kong probably has a higher concentration of testing functions than anywhere else in the area, with the exception of Singapore. All of the foreign-owned plants in the territory incorporate testing facilities (indeed all semiconductor plants do with the possible exception of one locally owned subcontractor), and three of them (Motorola, Fairchild, and Sprague) possess no assembly function, opting to concentrate on testing alone.

In addition to the emergence of Hong Kong as a regional testing centre, it is of particular interest to note that six foreign-owned companies (three US, two Japanese and one British) have set up

Table 5.11 Semiconductor manufacturers in Hong Kong, 1986

Company	*Country of ownership*	*Labour processes*	*Approximate employment*
Motorola	USA	d, t, r	750
National Semiconductor	USA	a, t	1,000
Fairchild	USA	t	100
Siliconix	USA	d[1], a, t	550
Teledyne	USA	a, t	200
Sprague	USA	t, r	200
Commodore	USA	a, t	200
Microsemiconductor	USA	a, t	100
Zilog	USA	d, r	100
Hitachi	Japan	d, a, t	200
Sanyo	Japan	a, t	200
Oki	Japan	d	20
Philips	Netherlands	a, t	500
Ferranti[1]	UK	d, a, t	500
Hua Ko	PRC	d, w, a, t	300
Elcap	Hong Kong	w, a, t	300
RCL	Hong Kong	w, a, t	200
Micro Electronics	Hong Kong	w, a, t	400
Semiconductor Devices	Hong Kong	a, t	900
Swire Technologies	Hong Kong	a, t	500
Century Electronics	Hong Kong	a	N/A

Key: d – design centre; a – assembly; r – regional headquarters; w – wafer fabrication; t – final testing

Sources: Interviews; Company Reports, Hong Kong Productivity Centre, *Electronics Bulletin* (various issues); *Global Electronics* (formerly *Global Electronics Information Newsletter*) (various issues); UNCTC (1986)

Notes: 1. Joint-venture with Hong Kong firm

customer-related design centres in the territory (in two cases, Zilog and Oki without also having manufacturing facilities), and as already mentioned, three US companies (Motorola, Sprague, and Zilog) now have their Asian regional headquarters there also.

Finally, at this point, we simply need to call attention to the emergence of a locally owned semiconductor industry currently consisting of seven firms. One of these is a fully owned subsidiary of the Chinese government's National Light Industries Corporation (*Asian Wall Street Journal Weekly*, 2 August 1982), and operates a design centre as well as having wafer fabrication capability. Three others fabricate, as well as assemble and test semiconductors.

The significance of these locally owned developments together with that of the other changes described above will be addressed in the next section. It is to that section and hence our analysis of the reasons why Hong Kong has emerged as a core of the regional division of labour that we now turn.

The making of a regional core

In Chapter two, I argued that when seeking to understand the emergence and operation of an industrial branch at a given location (or, indeed, economic development generally), it was important to recognise that those processes that were significant determinants originally may, and often do, alter and develop a different relative significance over time. Conversely, processes that had little or no significance initially, sometimes emerge and become important determinants of the future development of the phenomenon. Here, and in the next chapter on the semiconductor industry in Scotland, we will be concerned not merely with those processes that led to the emergence of the branch in Hong Kong or Scotland. We shall also be concerned with the way in which those processes have altered (or been formed) historically in the intervening twenty-five years, so as to give rise to a production system which (especially in the case of Hong Kong) plays a quite different role in the international division of labour than was initially the case.

Following one of the methodological leads which has guided our analysis throughout the study, we begin our account of the emergence of Hong Kong as a regional core by focusing on the processes that have been internally necessary to this development: those associated with both the production and realisation of surplus value in the semiconductor branch.

Markets

We begin with the influence of markets on the development of Hong Kong as a regional core of semiconductor production, not because we consider the realisation issue to be the central one with regard to the prospects for development. Indeed, in Chapter two, I explicitly argued against this conception in as far as it has held sway in the forms of dependency theory associated with Frank, Amin etc. Rather, the decision to begin with markets, in the Hong Kong case, stems more from convenience than it does from its significance in the theoretical system. By this, I mean that the nature and structure of markets, historically, have been of much less significance for the origin and development of semiconductor production in Hong Kong, than they have been, for instance, in Scotland (see Chapter 6). As a result, in the case of Hong Kong, they can be briefly dealt with.

As we shall see below and in Chapter seven, however, the question of markets may have greater import for the future development of semiconductor production in the territory, than it has had in the past.

The first point to emphasize here is that US semiconductor manufacturers did not invest in production facilities in Hong Kong in order to penetrate the local or regional (including Japanese) markets. Those markets then (1960s) were either almost non-existent, or, in the case of Japan, they were closed-off to foreign producers behind high tariff barriers (Okimoto *et al.* 1984; Grunwald and Flamm 1985; UNCTC 1986). Though regional markets have become more significant in recent years, particularly for re-exports, the United States and secondly the EEC were, and largely remain, the *raison d'être* in terms of markets, for both electronics, generally (see Table 5.8) and semiconductor production specifically, in the territory. In the case of IC exports in 1981, for instance, the United States and the EEC together constituted 79.6 per cent of Hong Kong's export market (derived from Table 4.3).

If the realisation of the surplus value generated by the territory's semiconductor industry has been dependent historically on markets at some geographical distance from the production site, what is likely to be the situation in the future? While regional markets have taken on a growing, if still limited significance for Hong Kong's semiconductor industry, of particular note is the increase in the proportion of sales to China (see Table 5.8 for electronics generally). At the present time the Chinese market is particularly important for the four locally/PRC-owned producers (Hau Ko, RCL, Elcap, and Micro Electronics). Should the Chinese government's plans for economic expansion and industrialisation be successful, then one would expect the Chinese market (which, after the transfer of power from the British colonial regime in 1997, will become formally Hong Kong's domestic market) to become increasingly significant for foreign as well as local producers in Hong Kong. The problem for the territory's semiconductor and electronics industries generally, however, particularly in the light of certain structural changes in the world economy (explored in Chapter 7), might be whether the Chinese economy can develop quickly enough to absorb an increasing proportion of the output. For the foreseeable future, for a variety of economic and political reasons, Chinese development must remain an open question.

Labour processes, labour power, the state, and related issues

If US semiconductor investment in Hong Kong was (and is) driven more by concerns with the production of surplus value than by its realisation, then we need to examine, and as far as possible assess, the relative causal weights of the various factors that brought semiconductor firms to the territory in the first place. In addition we need to assess the significance of changes in the original factors, and the emergence of new factors that have helped both maintain the presence, but alter the nature of the operation, of those firms that continue to produce in Hong Kong.

In this section we explore the inter-related sets of determinants that have emerged historically, both at the level of the world-system and internal to Hong Kong itself, and which have led to the territory over the past quarter century being perceived by semiconductor firms as a 'good place' to do business. Clearly many of the social, economic, and political features of Hong Kong that have encouraged investment in semiconductor production, are exactly the same as those that have encouraged investment in manufacturing industry generally (and indeed other industries also), and thus have helped to create Hong Kong as one of the world's most economically successful NICs. Although this is not the place to provide a 'total' account of the territory's economic development, it will be necessary in the subsequent paragraphs to discuss a number of the elements of such an account.[11]

At the time of initial semiconductor investment in the early 1960s, the legacy of Hong Kong's historical – indeed traditional role – in the world economy, meant that the territory possessed many of the contingent factors which NIDL theorists (perhaps especially Fröbel *et al.* 1980) regard as important for export-oriented industrialisation based on foreign investment. Hong Kong's traditional role as a British colonial entrepôt helped ensure that the territory was a free port. As a result the import and export of the materials necessary for the production of manufactured commodities were not subject to customs duties and no constraints were placed on the repatriation of profits. This freedom from state levies on surplus value may have been an important inducement to prospective industrial investors at a time when other East Asian countries did not possess the export processing/free trade zones (with their liberal fiscal arrangements) that they were subsequently to develop.[12]

Not only did Hong Kong's free port status mean that foreign investors could profit from the absence of fiscal and other state restrictions, but they could also take advantage of the relatively well-developed transport and communications infrastructure which again was usually part of the historical legacy of performing an entrepôt function. Some commentators (e.g. Reidel 1972) have mentioned another historically contingent factor – an English-speaking population – which they consider may have initially attracted foreign manufacturers to Hong Kong rather than elsewhere in the region. Unfortunately this factor cannot be given much weight in our explanation when one recognises that less than 25 per cent of Hong Kong's population speak any English at all (Bolton and Luke 1986), and that other parts of the region, such as the Philippines (which had been a US colony until 1946 and where about 85 per cent of the population spoke English) were, on this criterion, a much better bet.

It seems to me that other than the benefits which accrued to Hong Kong because of the legacy of its entrepôt function, there were few contingent factors which marked off the territory from its potential competitors in the region as a location particularly suitable for foreign-owned, export-oriented, commodity production.

What, then, are we to make of the locational determinants which arise from the social relations internal to manufacturing, and essential to it, as part of the process of capitalist commodity production? In Chapter two, I identified these 'internally necessary relations' as being sited within the valorisation process, and hence associated with the nature and organisation of the labour process and with the form and cost of labour power. From our earlier discussion, and drawing on some of the more 'orthodox' accounts of the internationalisation process, we know that the concern to reduce costs in the face of Japanese competition was perhaps the single most important reason why US semiconductor companies moved their assembly functions offshore. As we saw in the previous chapter, however, in the 1960s there was little to distinguish Hong Kong as a source of cheap unskilled labour, from those other parts of the region that were potentially sites for US investment. Indeed, if anything, wages in Hong Kong were possibly even slightly higher than many other parts of the region (see Table 4.1). What did set Hong Kong aside from its potential competitors, however, was not the cheapness of its labour *per se*, but a combination of relative cheapness with a particular 'quality' of labour. By quality I refer to the fact that by the early 1960s, Hong Kong, *alone amongst its Asian counterparts* (with the exception of Japan), had already produced a workforce which had been habituated to the social and personal demands of manufacturing labour processes for over ten years. As a result of the combination of world-system and geo-political factors discussed earlier in this chapter, Hong Kong, via the development of its textile industry, had gained a head start in the (newly) industrialising stakes.[13] As such, it could provide prospective manufacturers with a labour force that was well on the way to being culturally proletarianised, and hence one which was already socially adjusted to the regimes and rigours of factory labour.[14] Though I know of no empirical evidence to support this contention, it seems highly likely that Hong Kong, unlike its competitors, was able to deliver a workforce which both could be quickly trained for electronics assembly work, and, crucially, could be induced to operate effectively (without high rates of absenteeism, for instance) and productively within a relatively short period of time.[15]

If cost and quality of manual labour was perhaps the principal initial 'draw-card' which Hong Kong possessed *vis-à-vis* its neighbours, what were the factors which ensured the continuation of foreign and local investment, not only in semiconductors and electronics, but across the territory's manufacturing base? There seem to me to have been four inter-related factors. First, although labour costs rose consistently through to the early 1980s, they were maintained at levels below what they might otherwise have been, in part by indirect subsidies from both the Hong Kong and Chinese governments. Second, the quality of the labour power in the electronics industry was associated with an historic aversion to militant action by the local trade union movement. The third factor, which becomes particularly important for explanations of the recent technological upgrading of the semiconductor industry, was Hong Kong's ability to deliver highly skilled technical and engineering labour at a cost far below that which would be necessary in (say) the United States, Western Europe, or Japan. The final factor was the benefits (in terms of the local sourcing of components and materials, and the availability of skilled manpower) which accrued to foreign firms from the emergence of the local electronics and semiconductor production complex. The remainder of this chapter will be taken up with an assessment of the role of these factors in the emergence of Hong Kong as a regional core of semiconductor production.

Subsidizing labour costs

The ideologues of neo-classical economics have argued that the NICs of East Asia, and Hong Kong in particular, have been economically successful precisely because they have been relatively free from state interference with the 'normal' working of market processes (Friedman and Friedman 1981; Rabushka 1979). Hong Kong economists from the first general account of the territory's economic 'miracle' (Szczepanik 1958) to more recent expositions (e.g. Lin *et al.* 1980; Cheng 1985), have repeated these claims, as have the respective governments themselves (particularly the Hong Kong government with its persistent use of the ideologically loaded oxymoron, 'positive non-interventionism' in reference to its economic policies). Unfortunately for the economic orthodoxy, it is now becoming increasingly clear, that far from the East Asian gang of four being shining examples of what free-market economics can do, on closer inspection, they turn out, if anything, to be examples of state-led economic development. The 'truth' of this counter argument is now reasonably well accepted (except by the Chicago ideologues and their acolytes), at least as far as the development of South Korea, Taiwan (see Browett 1986; Harris 1987; Gold 1986; Deyo 1987 for general presentations of the argument), and Singapore (see Lim 1983) are concerned. The relationship between state policies and the economic development of Hong Kong, however, is less well known (including by political economists; see for instance, Harris 1987: 54–60).

While the Hong Kong government has not intervened directly in the capital market or in funding R & D, as the Taiwanese and South Korean governments, for instance, have done (Hamilton 1983; Schmitz 1984; Browett 1986; Harris 1987; Deyo 1987), it has intervened directly in the labour market and indirectly in both the capital and labour markets. Its intervention in the capital market has taken the form of very low corporate and personal taxation (both currently standing at 15.5 per cent maximum). These low rates of taxation (which remain lower than other states in the region; see BBDO 1987), given the Government's substantial expenditure on education and welfare (and policing!), have been achieved only because of its historic ability to generate a high proportion of its revenues from non-taxation sources. In fact, from 1949–50 to 1975–6, non-taxation revenues averaged over 34 per cent of total Government revenues (Schiffer 1983: 23). By far the largest source of non-taxation revenues has been the sale of leases on government-owned land, and in Hong Kong, the Government owns over 95 per cent of all land. Its subsidies to private capital, by means of low taxation, are then, in large measure, a result of the fact that perhaps unique amongst capitalist societies, land in Hong Kong is nationalised.

The Hong Kong government has intervened in the labour market directly, by means of its extensive educational provision at secondary and tertiary levels (and hence the 'extended' reproduction of labour power), but indirectly also, particularly by its contributions to the 'social wage'. These contributions have largely taken the form of welfare provision, and especially of low-cost housing. The Hong Kong government, since 1954, has developed the second-largest public housing system in the world (in terms of the proportion of the population housed – Singapore is the largest), which currently houses about 45 per cent of the total population, and over 80 per cent of the territory's working class (Castells 1986a; S. Y. Ho 1986). In addition, the Government organises what is, in effect, a cartel to control the prices of basic foodstuffs such as rice and vegetables.

Paradoxically, the Chinese government from the beginning (1949) has been involved in subsidising the wages of Hong Kong workers. It has done this by supplying the bulk of the territory's food requirements (rice, poultry, meat, vegetables) at 'administered' prices; that is, at prices significantly below those on the open market. In addition, it has provided cheap clothing (far below the cost of imported or domestically produced items) for the territory's working class via its numerous retail outlets in the territory. In an important, but as yet unpublished paper, Jonathan Schiffer has calculated that 'non-market' forces (i.e. the combined efforts of the Hong Kong and Chinese governments) subsidize the household expenditure of Hong Kong's working-class population to the tune of 50.2 per cent! (Schiffer 1983, Table 4: 19). Without such subsidies, manual wages would undoubtedly have risen faster in the territory than, in fact, has been the case.

Intervention in both capital and labour markets, in the forms that I have described, has had the effect of reducing production costs and thus acting as a significant inducement to investment by both foreign and local manufacturers. While it is in one sense difficult to distinguish the determinants of the cost of labour power from the determinants of its features when it is converted, via the labour process, into concrete labour, for analytic purposes this distinction must be made. Consequently, we now more to the question of how the quality of the labour force (that is, of concrete labour), has been reproduced over the years subsequent to initial semiconductor investment.

Reproducing the quality of labour

I argued in a previous section that the early social adaptation of Hong Kong workers to factory labour was probably a key factor in the initial attraction of US semiconductor (and other electronics) manufacturers to the territory. But the quality of Hong Kong labour was based not merely on the fact that workers could perform the routines of assembly work efficiently and quickly, but that they tended to do so with seemingly few expressions of resistance, of either an organised or individualistic kind.[16] That this situation has remained substantially true to the present day (with the exception of the anti-colonial riots of 1966–7), requires explanation.

While a comprehensive explanation of this phenomenon would have to take into account, among other things, the processes of legitimation and control within the colonial state and the mechanisms of ideological transmission, as well as the production of workers' consent within the workplace (cf. Burawyo 1979),[17] this is not the place to address those questions.[18] For the purposes of the current discussion, however, we focus on the role of trade unions and the composition of the labour force.

Though Hong Kong has had a trade union movement which has organised between 13.5 (in 1968) and 25.2 (in 1976) per cent of the employed labour force, and currently (1984) organises 16.1 per cent (Ng 1986a, Table 10.2: 276–7), it has not been one that has mobilised its members around the standard (for Western trade unions) issues of wages and working conditions. Neither has it adopted a militant stance to the workplace problems of its membership. Indeed, as we can see from Table 5.12, labour militancy gauged in terms of strike activity, though never high by European or American standards (or indeed by the standards of other East Asian countries, such as the Philippines), has declined substantially (in terms of working days lost) in recent years.

Table 5.12 Work stoppages in Hong Kong

Year	*Number of lock-outs and strikes*	*Working days lost*
1951–2	12	53,436
1956–7	12	78,852
1961	10	38,558
1966	12	24,355
1971	42	25,600
1972	46	41,834
1973	54	56,691
1974	19	10,708
1975	17	17,600
1976	15	4,751
1977	38	10,814
1978	51	30,927
1979	46	39,743
1980	37	21,069
1981	49	15,319
1982	34	17,960
1983	11	2,530
1984	11	3,122
1985	3	1,160
1986	9	4,908
1987	14	2,774

Sources: Commissioner for Labour, Departmental Annual Reports and Census and Statistics Department, *Monthly Digest of Statistics* Data for 1951–84 compiled by Ng (1986a: 272)

This situation of a lack of interest in workplace issues together with an absence of militant struggle is explained in the literature as being the result of the convergence of a variety of factors. Authoritative commentators (principally Turner, *et al.* 1980; England and Rear 1981) agree, for instance, that trade unions in the territory traditionally have been concerned predominantly with their role as welfare and 'political' agencies. In periods when the social wage was very limited, their welfare function clearly must have been of some significance for workers. Similarly, many Hong Kong unions, or at least those organised by the Federation of Trade Unions and the Trade Union Council, were forged during the political (and military) battles between the Communist Party and the Kuomintang. As a result they continue to put much of their energy and money into political propaganda and mobilization on behalf of China and Taiwan respectively.

Since the early 1970s, new labour organisations have emerged. While many of these have been formally constituted as unions, they have been independent of the two federations, and hence of their 'external' political concerns. These new unions, however, have developed predominantly among public sector employees, and as well as being relatively combative over wage issues, they have also been involved in mobilizing around non-workplace concerns such as utilities and public transport costs (Tso 1983; Ng 1986b).

Of greater interest for the current discussion, however, has been the emergence of a number of non-trade-union labour organisations. Born out of a combination of religious and student radicalism, the largest and most militant of these organisations has been the Hong Kong Christian Industrial Committee. While the CIC has been much more involved with workplace issues than most industrial unions, and has close contacts with new 'independent' unions such as the small (less than 1,000 members) Electronics Workers' Union, it has tended to concentrate the bulk of its energies on fighting for redundancy and maternity payments (often illegally denied by local employers) and educating workers about their legal rights, health and safety issues etc. (Chiu 1986).

Though not as directly repressive as in some parts of East and Southeast Asia (South Korea, Taiwan, Indonesia, the Philippines under Marcos, for instance), labour legislation has operated, additionally, to discourage workplace solidarity and mobilization. Trade union ordinances on the one hand allow minute sectional interests to gain formal expression (as few as seven workers are legally entitled to form a union) and on the other, they ban general unions. What is more, since as early as 1927 (in the immediate aftermath of the 1925–6 general strike), sympathy and 'political' strikes have been illegal (England and Rear 1981: 124–30).

In addition to the 'peculiar' features of the territory's trade union movement, commentators (e.g. England and Rear 1981) have pointed to the fact that since the late 1940s, the 'simple' reproduction of labour power in Hong Kong (in part) has taken the form of immigration from China. The argument has been that by virtue of the fact that the new arrivals have been escaping from political repression and poverty in China, their overriding concern in Hong Kong has been to provide materially for their families. In order to do that, they have been willing to work very hard, tolerate bad working conditions, long working weeks and exploitation generally. Unfortunately, when viewed from a comparative perspective, arguments about the supposed relation between immigrants and high levels of work discipline and docility, do not appear to carry much weight. The labour history of the United States, for instance, a country whose industrial dynamism was built in part upon wave after wave of immigrant labour, was also, at least prior to the Second World War, the history of one of the most combative working classes industrial capitalism has yet known (of the enormous literature see, for instance, Brecher 1973; Aronowitz 1974; Gutman 1977; Montgomery 1979; and in a more general context, Henderson and Cohen 1982b).

If immigration *per se* does not help us to account for the 'quality' of the Hong Kong workforce, is there something about the historical construction of Chinese culture which in articulation with the relations of power within the factory, and within the colonial state generally, helps to produce this effect? While we are unable to explore this hypothesis in any depth here, at least one (Hong Kong) Chinese scholar has argued that the 'deep structures' of Chinese culture tend to help reproduce, over time, a ready acceptance of authority and an abhorrence of face-to-face conflict (Sun 1983). If this is indeed the case, then it might take us some way to explaining the seemingly low levels of worker resistance to factory labour processes.[19]

One of the significant features of the electronics and semiconductor industries in Hong Kong, as in other parts of the world (cf. Lim 1982; Lin 1987, on Southeast Asia), is that an overwhelming majority of the workforce are women (see Table 5.4). Workplace ethnographies which have focused on women in wage labour in Western societies, have tended to argue that the patriarchal relations of domination into which women are socialised within the family, are transferred to the workplace where they are used creatively by managers and supervisors as a major component of the social basis of control (see, for instance, Downing 1981; Pollert 1981; West 1982). In the Hong Kong semiconductor industry, companies seemingly recruit women for assembly work precisely because they are 'dextrous and more willing to tolerate monotonous work' than are men.[20] Although there is no empirical work that deals specifically with the issue, it seems highly likely that systems of patriarchal control operate in Hong Kong factories as much as they do elsewhere. In addition, Janet Salaff's (1981) and Claire Chiang's (1984) work, shows that the principal life interests of female manual workers in the territory lie not within the experience of labour, but rather are heavily determined by familial obligations. They work, overwhelmingly it seems, because they are obliged to contribute to the household income.

The above comments are not meant to constitute an argument that women electronics workers in Hong Kong do not resist assembly labour processes. On the contrary, there is evidence that individualistic forms of resistance do occur (Djao 1976; Chiang 1984), and high worker-mobility ('labour-turnover') rates are endemic in the Hong Kong electronics and textile industries. As Cohen and I have argued elsewhere (Henderson and Cohen 1979), in the context of factory-based production systems, worker mobility, under certain conditions, can become an important form of resistance. The point, however, is that in the absence of research that specifically sets out to chart workplace resistance among Hong Kong women, we are left with the conclusion that what resistance there is, is less evident in both quantity and significance than it is, say, among their British couterparts (cf. Pollert 1981).

The semiconductor production complex

The previous discussion was an attempt to identify some of the principal reasons why US semiconductor manufacturers came to Hong Kong and remained there. Much of the weight of the argument was placed on the cost and quality of labour power and on the processes by which these were reproduced. In this section, we engage directly with the question of why US semiconductor firms have upgraded their operations, such that Hong Kong is now one of the technological cores of the East Asian regional division of labour. We begin, however, by recovering some of the earlier discussions.

In Chapter four, I suggested that until the late 1970s, US semiconductor production in Hong Kong was much the same as it was in other parts of the East Asian region. It was involved overwhelmingly, in other words, in the manual assembly of transistors, diodes, and integrated circuits. The data presented in Figure 3.2 (p. 47) confirmed this situation. I then went on to suggest that by the mid-1980s, the situation had changed drastically. In Hong Kong, as in Singapore (and to a lesser extent, South Korea and Taiwan) US companies had restructured their operations by engaging in a process of capital deepening and had begun, by and large, to specialize in more technologically advanced processes such as final testing and circuit design. In addition, a number of companies had established their regional headquarters in one or other of these locations. What assembly functions remained, tended to be automated. The data presented in Figure 4.2 (p. 56) confirmed the emergence of Hong Kong and Singapore as the technological and managerial cores of the regional division of labour, and Table 5.11 (p. 100) presented the Hong Kong situation in more detail.

As of late 1986 there were 21 companies in Hong Kong engaged, in some way, in the manufacture of semiconductors. Of these, eight were American, three Japanese, one Dutch, one British, one Chinese and six Hong Kong-owned. All the US companies in the territory have testing facilities. In earlier periods, when discretes and ICs were less technologically sophisticated than they are now, testing was a relatively simple process. In recent years it has become an automated, technologically complex labour process (using lasers and computers to identify faults). It requires substantial proportions of engineering and technical labour, though not insignificant amounts of unskilled labour. The latter group of workers are used to manually load the semiconductors into the channels which feed them into the testing equipment. Thus, Motorola's Hong Kong plant, probably the most advanced testing centre in the developing world, employs 750 workers, of whom only 300 are unskilled (and all of whom are women). Five of the US companies in the territory (National Semiconductor, Siliconix, Teledyne, Commodore, and Microsemiconductor) still engage in semiconductor assembly (Motorola has never assembled in Hong Kong). In all these cases, however, assembly is automated, to a greater or lesser extent, depending on the technological sophistication of the device itself. Typically, microprocessors and the more advanced memory ICs (which require many wires to be attached to them, and hence more 'bonding' operations) will be assembled automatically. This ensures not only greater productivity, but higher yields, and hence higher-quality products. Less advanced devices tend to be assembled semi-automatically.

Three of the US companies (Motorola, Siliconix, and Zilog) have design centres in Hong Kong. Motorola's design centre is one of its three largest outside of the United States, Siliconix's is a joint venture operation with an independent Hong Kong-owned design house (Central Systems Design), and Zilog's design centre is its only productive operation in the territory, and complements its Asian regional headquarters, which is also located in Hong Kong.

The other foreign companies producing in the territory, with the exception of the Japanese firm, Oki, have all invested in assembly and test facilities. In addition, two of the Japanese companies (Oki and Hitachi) have set up design centres, as has the single British producer in the territory, Ferranti. In the latter case, production has been developed on a joint-venture basis with the territory's largest subcontract assembler, Semiconductor Devices. Ferranti custom-design the circuitry in Hong Kong (but with a satellite link to the company's R & D centre in Britain); the masks are produced and the wafers fabricated in Britain and the ICs are assembled and tested in Hong Kong from whence they are distributed directly to the markets (Hong Kong Productivity Centre, *Electronics Bulletin* 3(1), 1983: 30–1).

Before I discuss the determinants of the upgrading of semiconductor production in Hong Kong, it is necessary to comment on the emergence of the locally owned presence in the industry.

There are now seven locally owned firms manufacturing semiconductors in Hong Kong. Of these, three are subcontract assembly houses. The largest of them, Semiconductor Devices, has advanced testing facilities and semi-automated assembly processes. As a result, it is able to assemble and test some of the most technologically sophisticated devices, including microprocessors. A fourth firm, Microelectronics, was, and remains a subcontract assembly, but it has recently moved into wafer fabrication also.

The other three firms emerged after 1982 with wafer-fabrication capabilities from the very beginning. One of them (Hua Ko) is wholly owned by the Chinese government, and the others (Elcap and RCL) are reputed to have some Chinese (PRC) equity participation. All of them have acquired equity participation in Californian design houses, though one of them (Hua Ko) has its own additional design capability. Both Hua Ko and RCL have had a number of their engineers trained in California (in the former case, engineers from the PRC, not Hong Kong).

These companies all have a similar operation. They design and have their masks produced in California, and they fabricate the wafers, assemble and test in Hong Kong. So far, they have been able only to produce the less-sophisticated devices, up to, and including, 64K memory ICs, and reputedly suffer from low yields, thus suggesting that they have quality control problems. While one of the companies (RCL) is trying to move into the more advanced (256K) memory production, and two of them (Hua Ko and RCL) are attempting to develop microprocessor capabilities, they remain, for the moment, technologically confined to the lower end of the market. Indeed, their principal customers are the Chinese consumer electronics industries and Hong Kong's watch industry (Hong Kong Productivity Centre, *Electronics Bulletin*, various issues).

Determinants of restructuring

We are now in a position to direct attention to the question of why semiconductor production in Hong Kong should have been restructured along the lines indicated above.

The first point to be made, is that unlike its counterparts in South Korea, Taiwan, and Singapore, the Hong Kong government has not been involved in an attempt to shift the industry from a regime of absolute to a regime of relative surplus value creation. There are no incentives nor pressures for manufacturers to technologically upgrade their operation. While the Hong Kong government supports a semiconductor design facility (as part of the Hong Kong Productivity Centre's Electronics Division), the amount of investment the facility receives is infinitesimal compared with current investment (equivalent of 3 per cent of sales) by the South Korean and Taiwanese governments, for instance.

If the restructuring of semiconductor production has not been a result of state development strategy, what else has it resulted from? While comments on the determinants are scattered throughout the previous discussion, we need to combine and highlight them here.

First, as I argued in Chapter four, Hong Kong, along with the other members of the gang of four, was successively 'priced-out' of the cheap labour market, beginning in the late 1960s (Table 4.1). In the case of Hong Kong, this was not the result of deliberate state intervention (as it was in Singapore; see Chapter 4), nor does it seem to have been associated with actual, or anticipated, labour militancy. We must qualify immediately, however, this last statement. Although official data are not collected, both employers (particularly in the electronics, textiles, and plastics industries) and authoritative commentators (Turner *et al.* 1980: Chapter 5; England and Rear 1981) consider the Hong Kong workforce to be highly mobile. Interviews with semiconductor industry executives confirm this picture, with some of them (Fairchild and National Semiconductor, for instance) reporting worker mobility (that is, 'turnover') rates in excess of 20 per cent per annum by the late 1970s. In the context of persistent labour shortages since the late 1950s (only partially remedied by waves of legal and illegal immigration), labour mobility by individual workers in search of a slightly better wage and a slightly more congenial labour process made considerable sense.

Under circumstances of rising labour costs, semiconductor (and other) manufacturers could presumably have relocated their entire production operations to other parts of the region where wages remained lower than in Hong Kong. As we saw in Chapter four, new investment in labour-intensive assembly processes was indeed directed to Malaysia, the Philippines and the like. But US producers, in large part, remained in Hong Kong, though on the basis of a technologically upgraded operation. A major reason for the retention of restructured production units in Hong Kong (and Singapore etc.), was that by the mid-to late 1970s, the Hong Kong education system could provide the qualified engineers and technicians who are the *sine qua non* of advanced testing facilities and design centres (not to mention wafer-fabrication plants), and in reliable quantities.

This does not mean to say that the Hong Kong tertiary education system can produce electronics engineers with innovative R & D capabilities. As with its equivalents elsewhere in the region, it probably cannot. But the most advanced semiconductor labour process currently implanted in the territory – circuit design – does not require fundamental innovative research capabilities. Like other labour processes, circuit design has been taylorised in recent years, such that much of the work involves the fairly routine adaptation of innovatory work to specific customer requirements. While this absence of an innovatory R & D capacity may not pose a problem, currently, for the Hong Kong, and other semiconductor industries in the region, it may have serious consequences for future development. We shall return to this issue in the penultimate chapter.

The retention of upgraded production facilities in Hong Kong by US semiconductor companies, does not merely result, of course, from the fact that the territory is a reliable source of engineering and technical labour. So much is clearly true to a far greater extent for the United States, Western Europe, and, indeed, some other developing societies, such as India. The factor which distinguishes Hong Kong (and Singapore, South Korea and Taiwan) from the former territorial units (though not from the latter), is that electronics engineering and technical labour can be provided there, not only reliably, but also far more cheaply than it can, in (for instance) the USA and Scotland. Thus in 1981, annual wages for electronics engineers in Hong Kong were 60 per cent lower than they were for their equivalents in the United States, three years earlier, in 1978. Similarly the wages for electronics technicians were 76 per cent lower than their US counterparts in the same time periods (derived from Table 5.14). It is partly for this reason that US, Japanese, and British companies have developed design centres in the territory, in addition to restructuring their productive operations so as to concentrate, to a significant extent, on testing functions.

The third reason why US companies have remained in Hong Kong on a restructured basis, stems from the emergence of the semiconductor (and the more general electronics) production complex itself. As I argued in Chapter three, the emergence of a production complex on the basis of both the horizontal and vertical disaggrega-

Table 5.13 Comparative annual wage rates: Electronics engineers and technicians (in US $*)

	USA[a]	*Scotland*[b]	*Hong Kong*[c]
Electronics Engineer	19,188	11,876	7,562
Electronics Technician	15,288	9,189	3,676

*The exchange rate for pound sterling and Hong Kong dollars against US dollar as at mid-1982 prices

Sources: Figures adapted from:
Troutman (1980), Locate in Scotland (1983), Hong Kong Productivity Centre, Salary Trends and Fringe Benefits in the Electronics Industry, 1981

Notes: (a) Figure for June 1978
(b) Figure for Spring 1982
(c) Figure for April 1981

tion of production units, tends, under certain circumstances, to result in those units becoming insistently locked into a particular spatial location. The reason for this is that important sets of specialised services (technical and otherwise) and supplies develop around the principal producers, and as a result transaction costs tend to be reduced. In addition, in an industry as based on knowledge as semiconductor production is, regular interaction between executives and engineers becomes an important mechanism whereby new ideas and market information are acquired. In as far as it is important to one's operation to know what developments are taking place within Asian semiconductor production, there is no point in having one's most advanced functions (design, for instance) in the Philippines or Thailand. There is every point, however, in locating them in Singapore, Taiwan, South Korea, or Hong Kong.

Once a company has implanted production facilities in a particular territorial unit, and once a production complex begins to develop and deepen in that location, then for the reasons indicated above (assuming continuing access to major markets) there may well be compelling economic, social, and technical reasons for continuing to invest and upgrade production there, rather than at other, potentially competitive locations.

In Chapter two, I suggested that problems associated with the realisation of surplus value were, in general, a primary determinant of the globalisation of semiconductor production. Earlier in this chapter, however, I argued that in the case of Hong Kong, realisation questions (in the sense of local or regional markets) were not a major consideration for US companies who initially invested in the territory. The increasing linkage of the Hong Kong to the Chinese economy, however, may be changing this situation. While it may mean that what remains of low value-added production in the territory, could well be shifted to China in the next few years to take advantage of cheap labour there, (this is definitely the intention of National Semiconductor, the China connection could well have positive consequences for increased investment in higher value-added production processes. Although developments such as this would have negative consequences for manual workers in Hong Kong, it would probably mean that the territory would become the technological (and presumably managerial) core of a specifically Chinese 'national' division of labour in semiconductor production.

While a whole series of questions about future prospects envelop the electronics and semiconductor industries of Hong Kong and elsewhere in the region, we shall delay our assessment of them until Chapter seven. At this point in the study we must turn to an examination of the European role in the industry's international division of labour, and specifically to the case of Scotland.

Chapter six

Scotland: The European connection

In the preceding two chapters I argued that while US semiconductor companies may initially have invested in East Asian locations in order primarily to tap their supplies of cheap unskilled labour, the subsequent development of the industry in that region, and particularly in the gang of four, cannot be explained in such a simplistic way. In this chapter, we turn to our European case study, Scotland. From Figures 3.2 (p. 46) and 4.2 (p. 56) it is clear that the more technologically advanced labour processes, and particularly wafer fabrication, are the ones which US companies have established in Europe. What the data in those figures immediately indicate, then, is that cheap unskilled labour was not one of the reasons why US companies set up plants in Europe. Though, as we shall discover later, the cheapness and quality of engineering and technical labour (particularly in Scotland) may well be a primary reason why US (and other) companies continue to invest in advanced labour processes in Europe, the initial impetus for investment there was more the existence of a substantial and growing market, but one protected by high EEC tariff barriers.

From Table 1.1 (p. 7) it is clear that Britain has the fourth largest semiconductor industry in the world after the US and Japanese giants and just behind the European leader, West Germany. While a small amount of British semiconductor output is produced by indigenous companies such as GEC, Plessey, Ferranti, and INMOS, by far the most important contribution has resulted from investment by foreign and especially American companies. The preferred location for foreign semiconductor investment has been the central belt of Scotland. From its beginnings in 1960 with investment by the Microelectronics Division of the Hughes Aircraft Corporation, Scotland has developed to become the principal location for semiconductor production, not only in Britain, but in Western Europe. By 1983, Scotland had become responsible for 79 per cent of British and 21 per cent of European integrated circuit production (Locate in Scotland, 1983, Table I, 12).[1]

The chapter begins by analysing the reasons for the emergence and subsequent development of semiconductor production in Scotland. It then moves on to assess the role of Scotland in the US semiconductor industry's international division of labour. The significance and likely trajectory of the industry in Scotland, in the light of various tendencies in semiconductor production at the global level, will be assessed along with the possibilities for the various parts of East Asia, in the next chapter.

American semiconductor production in Scotland[2]

The electronics industry in Scotland originated with the British defence contractor, Ferranti, who set up a plant near Edinburgh in 1943. The first American electronics presence began with investment by NCR, Honeywell, and Burroughs in the late 1940s, and IBM in 1951. In each case these firms set up plants to assemble their various products for the British and European markets using imported American components (Campbell 1980; Crawford 1984).

The semiconductor industry in Scotland dates from 1960 when the Hughes Aircraft Corporation set up a diode assembly and test plant in the new town of Glenrothes. They were followed in the late 1960s by Motorola, National Semiconductor, and General Instrument, and in the early 1980s by the small defence contractor, Burr-Brown (see Figure 4.2), and the major Japanese producer, Nippon Electric (NEC). This foreign presence in semiconductor production has recently been supplemented by the emergence of an indigenous producer, Integrated Power Semiconductors.

The timing of American semiconductor investment in Scotland roughly coincides with similar investment in other 'offshore' locations such as Latin America, and particularly, after 1961, East Asia (Chapter 4). The labour processes that have been implanted in Scotland, however, are very different from those evident in Third World locations. Although initial investment in Scotland was in warehousing, test facilities, and occasionally assembly, it seems to have been the case that such limited investment was largely a means of 'testing the water', for within a short time of establishing a presence (two to four years), all the American firms had invested in the more capital-intensive, technologically sophisticated process, wafer fabrication. No American firm prior to 1987 had invested in offshore wafer-fabrication facilities other than in Europe, Japan, or Israel and thus their East Asian plants, as we have seen, remain substantially dependent on assembly processes, or in certain locations (Hong Kong and Singapore in particular) on final testing and design functions (Figure 4.2).

While the details of the semiconductor labour processes that have been implanted in Scotland, and issues related to them, will be taken up later in the chapter, it is first necessary to convey some of the general features of semiconductor production in relation to the electronics industry as a whole.

Electronics and semiconductors in the Scottish economy

Though underpinned by a consistent record of technological innovation for over a century and a half, the electronics industry in Scotland has been predominantly a post-war phenomenon. Indeed, as Table 6.1 indicates, it has been only since the early 1970s that electronics has made a significant contribution to manufacturing employment in the country. What is more, its increasing relative significance as an employer during the last ten years or so, is in part due to the rapid decline in employment opportunities in the more traditional manufacturing sectors (steel, shipbuilding, coal, etc.).

Table 6.1 Employment in the Scottish electronics industry, 1960–84* (000s)

	1960	*1970*	*1978*	*1979*	*1980*	*1981*	*1982*	*1983*	*1984*
Total manufacturing employment in Scotland[(a)]	733	709	609	603	566	502	478	447	437
Electronics employment in Scotland [(b)]	7	50	39	42	43	41	40	43	45
Electronics as a percentage of manufacturing employment	1.0	7.1	6.4	7.0	7.6	8.2	8.4	9.6	10.3

*Figures rounded to nearest 1,000.

Sources: (a) 1960–81: Census of Employment; 1982–84: Regional Data System, Industry Department for Scotland. Adapted from MacInnes and Sproull (1986, Table 3.9: 21)

(b) 1960: Firn and Roberts (1984: 298). Figure is for 1959; 1970: Scottish Manufacturing Establishments Record (SCOMER). Figure provided by John MacInnes; 1978–84: Adapted from Industry Department of Scotland *Statistical Bulletin*, C2.1, (September 1986, Table 1: 3)

Though electronics as a whole now accounts for over 10 per cent of manufacturing employment, the semiconductor industry itself, of course, makes a much smaller contribution.

Table 6.2 Employment in the Scottish electronics industry by product area, 1978–85* (000s)

Product area	*1978*	*1979*	*1980*	*1981*	*1982*	*1983*	*1984*	*1985*
Electronic data processing equipment[(a)]	6.3	6.7	7.2	6.8	6.4	8.7	9.1	9.7
Electronic components[(b)]	10.7	11.0	10.0	8.2	8.3	8.6	10.1	9.3
Electronic instrument engineering[(c)]	5.0	5.3	5.8	5.3	5.0	4.9	4.8	5.0
Consumer products[(d)]	0.6	0.5	0.6	0.4	0.5	0.5	0.7	1.0
Other electronics[(e)]	15.7	16.7	18.3	18.5	18.2	18.7	18.5	17.4
Totals	38.3	40.2	41.9	39.2	38.4	41.4	43.2	42.4

*Figures rounded to nearest 100

Sources: Regional Data System, Industry Department of Scotland. Complied by Fiona Deuchars

Notes: (a) 1980 Standard Industrial Classification (SIC) Activity Heading 3302
(b) 1980 SIC Activity Headings 3444, 3453
(c) 1980 SIC Activity Headings 3710, 3732
(d) 1980 SIC Activity Headings 3454
(e) 1980 SIC Activity Headings 3433, 3441, 3442, 3443

As Table 6.2 indicates, electronic components, of which semiconductors are a sub-section, have accounted for an annual average of about 9,500 jobs in recent years, or by 1985 a little under 22 per cent of employment in the electronics industry. By that same year, semiconductors provided about 4,400 jobs, or a little over 10 per cent of electronics employment (Table 6.3) but no more than 1 per cent of total manufacturing employment in Scotland. What is more, the semiconductor industry has consistently failed to match the expect-

Table 6.3 Semiconductor employment in Scotland, 1985

Company ownership	*Number*	*Employment*	*Proportion of total employment*
			%
USA[(a)]	5	4,060	92.5
Japan[(b)]	1	230	5.2
Scotland[(c)]	1	100	2.3
Total	7	4,390	100.0

Sources: (a) Company data
(b) Baggott (1985a: 18)
(c) estimate

ations for job opportunities that it has projected, and which have been echoed by the Scottish Development Agency. The 7,000 jobs by 1985 announced by the Agency (Locate in Scotland, 1983) have never materialised, and in the light of the industry's deepening crisis (Ernst 1987) seem unlikely to do so in the foreseeable future.[3]

The semiconductor industry, then, has had an abysmally poor record in terms of providing employment for Scottish workers. It has, however, had a better record in certain skill categories than others, but this is something we shall return to as part of our discussion of labour force composition. At this point it is important to recognise, along with other industry analysts (e.g. Firn and Roberts 1984), that the significance of semiconductors to the Scottish economy, far outstrips the industry's limited provision of employment opportunities.

First, as Table 6.4 indicates, while manufacturing output in Scotland has remained roughly stable in recent years, electronics output has risen by around 120 per cent in real terms, while the output of components (including semiconductors) has risen by about 50 per cent.

Second, semiconductors are the technological core of the entire electronics industry, and it would be difficult to imagine electronics as a principal basis for a revitalised Scottish industrial economy without a significant semiconductor-producing capacity. Herein lies perhaps the key reason why ownership and control over the production base, and hence the types of labour processes and technologies that have been implanted by semiconductor companies in Scotland, as well as the way these features of the industry have altered in relation to the changing international division of labour, become so important.

American penetration

By mid-1985, of 275 electronics companies manufacturing in Scotland, 42 of them (15 per cent) were American-owned. Of the rest, 178 (65 per cent) were Scottish-owned, 49 (18 per cent) UK-owned other than Scottish, with far smaller numbers owned by other European and Japanese corporations.[4] Though not numerically dominant at the level of the firm, American companies employed 41 per cent of the total electronics workforce, as did also their UK-owned equivalents. Scottish companies employed only 17 per cent of the electronics workforce.[5] If, on the basis of these indices, we can conclude that Scotland's electronics industry is significantly penetrated and controlled by American industrial capital, what of the situation with semiconductors?

Table 6.4 Electronics and all manufacturing output: Scotland, 1978–84 (Index values, 1980 = 100)

	1978	*1979*	*1980*	*1981*	*1982*	*1983*	*1984*
All manufacturing[(a)]	107	107	100	97	97	96	101
All electronics[(b)]	83	89	100	102	115	143	187
Electronic components[(b)]	92	99	100	93	100	106	140

Sources: (a) MacInnes and Sproull (1986, Table 3.9: 21)
(b) Industry Department of Scotland *Statistical Bulletin*, C1.1 (January 1986, Table 6: 6). 'Electronic Components' refers to 1980 SIC Activity Headings 3444, 3453

As can be seen from Table 6.3, the semiconductor industry in Scotland is totally dominated by American corporations both in terms of the numbers of firms and the proportion of the employment opportunities which they provide. What is more, two American firms, Motorola and National Semiconductor, can themselves be said to dominate the industry insofar as together they currently account for nearly 3,000 jobs, or 65 per cent of total employment (Table 6.5). Though for reasons indicated later, we can expect an expansion of employment opportunities in some of the other companies within the branch (Hughes and NEC in particular), and a contraction in Motorola and National Semiconductor, for the moment it appears that what is good for these two companies is good for 'Scotland's' semiconductor industry.

Having surveyed a few of the general features of the industry, at this point it is necessary to examine some of the determinants of the industry's development. This is done, first in relation to Scotland

Table 6.5 Engineering and technical labour at selected American semiconductor plants in Scotland, 1985

Company	*Total workforce*	*Engineers/ Technicians*	*Proportion*
			%
Motorola	1,750	400	23
National Semiconductor	1,100	275	25
General Instrument	210	70	33
Hughes Aircraft	800	240	30
Total	3,860	985	26

Source: Company Data

itself, before analysing in the subsequent section, Scottish semiconductors in the international divison of labour.

Determinants and development of semiconductor production in Scotland

I argued in Chapter two that when seeking to understand the processes that have determined the operation of an industrial branch in a given location, it was important to recognise that those processes that may have had significance originally, often alter and develop a different relative significance over time. In addition, processes that barely existed at the moment of the initial investment in production facilities, come to take on a determining role in the future development of those facilities. Though the need to provide a temporal dimension to analyses of industrial development and its impact in given locations, may seem an obvious point to make, it is an element which has often been missing both from traditional locational analysis (cf. Massey 1984) and NIDL theory (see Chapter 2).

In our efforts to analyse the development of semiconductor production in Scotland, as with our analysis of the East Asian situation, we are concerned not just with the factors which brought the industry to Scotland (and indeed, particular parts of Scotland) in the first place, but also with the way in which those factors have changed, or emerged, over time. We begin, however, by confronting the question of why it should have been Scotland, rather than any other part of Britain (or indeed Europe), which became the preferred location for American semiconductor investment.

I have argued consistently throughout this study that if we seek to generate theoretically coherent and plausible explanations of locational decisions we need to take seriously the realist distinction between the internal conditions necessary for the existence of a phenomenon and the contingent factors which help to determine the particular empirical form that the phenomenon takes. I suggested (in Chapter 2) that with regard to the spatial arrangement of any industrial branch, it is the fact that the branch is part of capitalist commodity production at large, that identifies the site of determinants internal to the branch as both the valorisation and circulation (of the commodity) processes. It is in this sense that calculations associated with the nature of the labour process (capital-labour ratios, labour costs, labour militancy, or passivity etc.) and with the structure of the market (competition, customer requirements, etc.) come to take on a crucial significance while the state often takes on a central, if contingent role. In contrast to other recent work on the determinants of high technology industry in Scotland (e.g. Haug 1986), this section, as in our previous case studies, seeks to separate out the internally necessary factors from the various contingent relations, which in articulation with one another, have given American semiconductor production in Scotland its current shape.

We begin by examining the role of markets, before turning to the question of semiconductor labour processes in Scotland and the various economic and social issues associated with them.

European markets

Though overtaken by Japan in the early 1980s as the world's second largest semiconductor market after the United States, in the mid-1980s, Western Europe still represented in excess of 17 per cent of world consumption (Table 6.6). In terms of national markets, Britain was the second most important after West Germany with 20 and 33 per cent of the European market respectively in 1982. Furthermore, together with France, the British semiconductor market was the fastest growing with about a 13 per cent growth rate between 1977–82 (UNCTC, 1986, Table II.9: 25). Without the emergence of markets for semiconductors in Britain and Europe, it is hard to imagine that American companies would have set up plants in Scotland, or anywhere else in the continent for that matter. But it was not simply the fact that there were substantial sales opportunities that brought semiconductor houses to Scotland. It was rather the structure of the market in relation to the nature of the product, together with a number of juridico-political factors which ensured that the companies manufactured, rather than merely marketed from bases in Britain or Europe.

Table 6.6 World market for semiconductors, 1980–5 (percentage by value)

Region/Country	*1980*	*1981*	*1982*	*1983*	*1984*	*1985*
USA	50.1	54.9	53.3	52.2	52.0	52.3
Japan	19.3	19.7	20.5	21.1	17.3	19.9
Western Europe	21.7	17.5	17.5	17.3	20.1	17.2
Rest of the world	8.2	7.9	8.7	9.3	10.1	10.6
Total	99.3	100.0	100.0	99.9	99.5	100.0

Source: Compiled from UNCTC (1986, Table II.6: 23)

Of significance here is the market generated by state purchases. Many of the semiconductor houses in Scotland produce a low volume, high value, customised product, much of which is directed to

British and European defence manufacturers (the principal exceptions being Motorola, and to a lesser extent, National Semiconductor). Because of stringent quality control requirements which defence manufacturers demand of suppliers of electronic components, and the systematic and regular customer-supplier liaison that this implies, it has become increasingly difficult for a components manufacturer to break into a national defence market without producing at least the technological core of the component within the particular national or regional boundaries themselves. It is no coincidence, therefore, that the first semiconductor house to establish a manufacturing plant in Scotland was Hughes Aircraft, a company which then (1960) and now, predominantly supplies the defence market.

If the need to manufacture adjacent to major British and European customers was an important reason for American semiconductor firms setting up British plants in the first place, then the additional need to penetrate EEC tariff barriers and hence compete effectively with European suppliers was a significant market-related determinant for those companies, such as Motorola and National Semiconductor, that began to set up operations in the early 1970s. The fact that the 17 per cent EEC tariff was levied on the value added during the production process, meant that semiconductor firms could produce most of the valued-added in their European plants – namely at the wafer fabrication stage – and cheaply assemble and test the semiconductors in their East Asian facilities, then subsequently import the completed product back into the EEC.

Throughout the 1960s and 1970s, most of the semiconductor companies in Scotland were fabricating various types of wafers which in effect duplicated for the European market, wafers which they also fabricated in the United States, for the American and other markets. While this form of the global organisation of wafer fabrication still exists for those companies which supply the small batch, customised markets – General Instrument, Hughes and Burr-Brown – there has been a significant alteration for the manufacturers of the large batch, standardised products, Motorola and National Semiconductor. While these plants certainly retain their European remit for certain types of semiconductors, their wafer-fabrication capacity has been increasingly given over to specialisation in one type of wafer for world-wide, and not simply just European markets. The implication of this, then, would seem to be that while the customised producers remain dependent on specifically European markets for the continuation of their Scottish operations, this is not necessarily the case with the large volume standardised producers. While the implications of this shift in market orientation will be examined in the next chapter, it is necessary at this point to turn to the second set of internal determinants, those associated with problems of valorisation, and hence with the particular labour processes that American semiconductor producers have established in Scotland.

Labour processes and related issues

I suggested earlier that of the five semiconductor labour processes, it has been wafer fabrication that has been predominantly implanted in Scotland by American companies. This is not to deny, however, that automated assembly and testing processes are likely to become an important feature in the future, or that some limited R & D (in the form of customer-related circuit design, rather than 'basic' research) has existed in the past,[6] but these are issues which will be raised in a subsequent section. Wafer fabrication is, for the most part, a highly capital-intensive labour process. The labour power it does utilise, however, tends to be that of highly skilled electronics engineers and technicians, though significant numbers (given the current state of fabrication technology) of semi- and unskilled workers are also required.

As can be seen from Table 6.5, about 26 per cent of the current semiconductor workforce in American-owned plants are engineers or technicians. If we bear in mind that some proportion of the non-technical labour force are managerial, supervisory, or clerical staff (450, or a further 26 per cent at Motorola, for instance), then we have a situation where probably no more than 50 per cent of Scotland's semiconductor labour force are manual workers. All the signs are, that with increasing automation, this proportion is likely to fall in the coming years.[7]

With this relatively heavy reliance on engineering and technical labour, a crucial determinant of the locational decisions of firms seeking to establish wafer-fabrication plants are (a) the availability and cost of such labour power, (b) the existence of facilities necessary for its reproduction and (c) some reasonable guarantee of a conflict-free industrial environment.

NIDL theorists such as Fröbel, *et al.* (1980) have stressed the availability of cheap labour in peripheral economies as a determinant of the locational decisions of multinational corporations. While our discussion in Chapters four and five has shown that this argument is simplistic for semiconductor production in East Asia it does have a certain credence in the Scottish context. While Fröbel *et al.* are concerned with the cheapness of manual labour (as labour processes implanted in Third World societies have tended to have low capital-labour ratios), this is not an especially significant factor for labour processes as capital-intensive as wafer fabrication. What is becoming increasingly important here, however, as with high technology industry generally (Castells 1986b, 1988) is the relative cheapness (and availability) of skilled, technical labour power. Wage rates for semiconductor engineers in Scotland in recent years have been, for instance, at most 60 per cent of those commanded by their counterparts in California (derived from Table 5.14). Furthermore, wage rates for engineers in Scotland have been consistently below those of engineers in other EEC countries such as France, West Germany, and the Netherlands (Locate in Scotland, 1983).

These remarkably low wage rates for this, the most important form of labour power necessary for wafer fabrication, are complemented by a steady supply from Scotland's highly developed university and technical college system. As mentioned previously, Scotland historically has been a major location for technological innovation and this tradition continues to be deeply rooted in the electronics departments of its universities (particularly Glasgow, Edinburgh, and Herriot-Watt). Scottish universities and colleges have continued to expand their output of electronics engineers and technicians and a number of the heads of American semiconductor firms in Scotland now consider their products to be among the best in the world.

While accepting the importance of engineers and technicians for the particular semiconductor labour processes that exist in Scotland, it is the case that those labour processes would not be viable without a significant input from semi- or unskilled workers. Though the cost of such labour power does not carry the relative weight as a determinant of locational and investment decisions as the cost of its more skilled counterparts, labour costs of a more general nature retain their significance. By this I have in mind those costs associated with the reproduction of the various forms of labour power. As a result of massive state investment in health, social welfare, and educational provision, Britain (in spite of recent public expenditure cuts) has the lowest reproduction costs to be borne by employers of any country in Western Europe (Table 6.7).

Though the cost of unskilled labour power may not be an important determinant of semiconductor investment and the subsequent development of the industry in Scotland, the 'quality' of that labour power undoubtedly is. By 'quality' I refer to the extent to which workers are reliable, efficient, and particularly, have a low propensity for industrial conflict. Whereas the high propensity for conflict traditionally associated with the Scottish working class might have been considered a negative feature for semiconductor firms thinking of establishing plants in the country, this problem has

Table 6.7 Employee benefit costs met by employers as a proportion of total earnings (selected EEC countries, 1984)

Country	*Employee Benefit Costs* (percentage)
Italy	52
France	45
West Germany	43
Britain	23

Source: SDA Data, reported in Walker (1987)

been successfully circumvented (at least for the moment) in the following ways:

1. The companies deliberately recruited women rather than men. As a result they have been able to utilise the sense of discipline that women have acquired through subjection to patriarchal domination in the household.[8] In this sense the companies have merely reproduced the employment pattern evident in their US and Third World plants (Siegel and Markoff 1985; Lin 1987). Thus 74 per cent of manual workers in the electronics industry in Britain's 'assisted areas' (including Scotland) are women (Cooke, *et al.* 1984: 284; see also Goldstein 1984; McKenna 1984).

   Not only has it been women *per se* who have been recruited, however, but young women in particular. The logic of such selective recruitment is that young people have not yet had the chance to adopt 'bad' work habits (including familiarity with trade union practices) and hence can be readily habituated to the labour processes, routines and expectations of the particular company. Nippon Electric (NEC) seem to have taken this strategy to its logical conclusion. In their Scottish plant they recruit only manual workers who have had no previous work experience. As a result, the average age of their manual workforce is reputed to be as low as 17 years old (Baggott 1985a).[9]
2. American semiconductor, and electronics firms more generally, have effectively warded off the unionisation of their Scottish plants. While many of the companies employ anti-union policies throughout their global operation (though not the computer manufacturers, Burroughs and Honeywell, whose Scottish plants are unionised), it is possible that IBM, the principal US electronics firm in Scotland, has set the tone. IBM has, apparently, consistently put pressure on the small companies, not

to recognise trade unions and has provided advice on how to resist unionisation.[10]

Probably more important than any actual resistance to unionisation, however, has been the fact that US plants tend to combine better-than-average wage rates and bonus systems with 'creative' and relatively egalitarian industrial relations systems and generous benefit packages. While the computer manufacturer, Hewlett-Packard probably provides the best example of creative industrial relations in Scotland (Cressey 1984; Cressey, *et al.* 1986), Hughes, with for instance, its egalitarian and generous annual leave system, may be the best example among the semiconductor houses.[11] The result of these techniques has been that not one semiconductor plant in Scotland has ever been unionised (MacInnes and Sproull 1986).

3. Semiconductor firms have attempted to secure low levels of conflict by adopting a particular spatial structure for their operations. Rather than locate their plants within the spatial products of an earlier round of industrialisation, they have overwhelmingly opted for the new towns of East Kilbride (Motorola), Livingston (Burr-Brown, as well as NEC and Integrated Power Semiconductors) and Glenrothes (General Instrument and Hughes). In fact, over 50 per cent of electronics investment in Scotland went into these three new towns during the period 1980–4 (Walker 1987). Of the semiconductor companies, only National Semiconductor have opted for one of the older industrial towns (Greenock). By opting for the most part for new towns which are relatively separate in spatial terms from the older industrial centres, the semiconductor companies have tried to avoid 'contamination' of their workforces by the traditionally militant socialism, particularly of the Strathclyde region. By means of this spatial solution, they have managed to take advantage of what are, in effect, isolated local labour markets from which they draw the bulk of their manual labour forces. In the case of Motorola, for instance, 60 per cent of their workforce live within a three-mile radius of the plant, and as a result of the family and friendship networks usually associated with local labour markets, they never have to advertise for manual workers.

So effective have these recruitment, habituation and reproduction processes seemingly been, that, in the context of high unemployment levels, which by mid-1985 had reached more than 16 per cent in those parts of Scotland where semiconductor plants are located (Fraser of Allander Institute 1985: 52), industrial conflict, for the moment, seems substantially to have been contained.

If these, then, have been the principal determinants internal to semiconductor production that have been responsible for the location and subsequent development of American plants in Scotland, what then have been the contingent factors which have helped ensure their presence there? While the usual factors such as good telecommunications, transportation networks and airport facilities, the availability of factory space, pleasant environments seemingly necessary for the social reproduction of high-tech executives (Hall 1985), are all as evident in Scotland as in many other parts of Britain and Europe, there are two contingencies that set Scotland apart as a prime location for semiconductor production. These are (a) state policy in terms of a system of regional aid together with an institutional apparatus designed to reduce expenditure and technically facilitate factory location; and (b) the emergence of a local electronics complex within which semiconductor production can thrive.

State regional policy

In previous chapters I have mentioned the ways in which national states operate to effectively reduce the costs of semiconductor (and other) manufacturers. In this chapter I have already suggested that state-generated defence markets have been important, while welfare expenditure by the British government has subsidized reproduction costs more substantially than elsewhere in Western Europe. An additional way in which state intervention helps reduce costs for semiconductor/electronics manufacturers in Scotland is through the mechanism of regional aid.

Until late 1984, most of northern Britain, including practically all of Scotland was designated by the British government as 'assisted' areas. Amongst the various forms of financial aid which could accrue to firms seeking to locate plants in those areas, the grant of 22 per cent of capital costs (now cut to 15 per cent on new investment only) was particularly important for a productive process as capital-intensive as wafer fabrication (Young and Hood 1984). Though the subsequent restructuring of regional aid to place more emphasis on employment creation may have helped divert some semiconductor investment away from Scotland,[12] the existence of such grants undoubtedly helped 'swing-the-balance' in favour of the country in the early years of its emergence as a semiconductor complex. The point does need to be emphasized, however, that state financial aid, generally speaking, is very much a secondary determinant of the locational and investment decisions of semiconductor manufacturers (Cooke, *et al.* 1984). In this context, it is of interest that inward electronics investment in Scotland in the 1980s has occurred despite the fact that regional development grants are now half of what they were during the earlier investment wave of the late 1960s (Walker 1987).

In addition to these financial subsidies, the activities of various central and local state agencies have been minor determinants of semiconductor production in Scotland. The centralisation of the administration of financial aid, the targeting of potential investors, the provision of various forms of technical assistance etc. in which the Scottish Development Agency and its division Locate in Scotland engaged, are especially important. Furthermore, the activities of 'development corporations' in the new towns has eased the administrative burden (in terms of the labyrinth of planning, environmental, health and safety concerns and taxation law), hence helping to reduce the 'start-up' costs of semiconductor manufacturers. Motorola in particular seem to have gained substantial 'hidden' subsidies of this sort from the East Kilbride Development Corporation.

The Scottish semiconductor production complex

In Chapters four and five, I argued that the emergence of local semiconductor production complexes in a small number of East Asian locations has helped to ensure that capital-intensive labour processes and more sophisticated production technologies are likely to be established there by American companies, rather than elsewhere within the Asian division of labour. Similarly, whatever the determinants of the initial investments by US companies in Scotland, the emergence of a strong semiconductor production complex there, represents an important inducement for subsequent investment.

The early establishment of generalised electronics production associated in part with US corporations such as Honeywell, Burroughs, and particularly IBM, helped provide both a ready market for semiconductors and symbolic re-assurance, by virtue of their presence there, that Scotland was a good place to manufacture electronic products. With the concentration of American semiconductor plants in Scotland, other foreign producers have been drawn in (NEC of Japan) and locally owned firms stimulated (Integrated Power Semiconductors). The international significance of the semiconductor complex, has subsequently attracted foreign-owned suppliers and sub-contractors and stimulated locally owned ones. Especially significant among the former have been the Japanese silicon wafer manufacturer Shin-etsu Handotai, and the American assembly and test sub-contractor, Indy Electronics. Among the latter, the mask-producing firm, Compugraphics and six design houses are of particular significance, technologically. With the exception of fabrication equipment, Scotland now possesses practically a full range of suppliers manufacturing for, or servicing, the industry's needs.

Table 6.8 Semiconductor support industries in Scotland

	Number of firms by location of ownership					
Production area	*Scotland*	*UK*	*Rest of Europe*	*USA*	*Japan*	*Totals*
Silicon wafers					1	1
Circuit designs	5			1		6
Masks	1					1
Assembly and test				1		1
Quartzware			1	2		3
Chemicals (specialist)	3		1	2		6
Metals (specialist)		1		2		3
Gases (specialist)		2				2
Packaging ceramics and metal stamping	1			3		4
Assembly equipment			1			1
Burn-in (specialist)	1	1				2
Production equipment	4			1		5
Maintenance (specialist)	1					1
Pipework (specialist)	1					1
Totals	17	4	3	12	1	37

Sources: Locate in Scotland (1983); SDA (1984a, 1984b), Baggott (1985b)

A scan of Scotland's semiconductor complex suggests that of a sample of firms whose products or services are directly related to the semiconductor industry, 17 of the 37 firms sampled (46 per cent) are Scottish-owned, while a further 12 (32 per cent) are US-owned (Table 6.8).

While it is not possible, at this stage, to provide data on

employment in the semiconductor-support industry, we need to recognise that most of these firms are highly capital-intensive and hence are likely to employ small numbers of mostly highly skilled (and we might add, male) engineers and technicians. Furthermore, when we chart the spatial arrangement of the Scottish semiconductor complex (Figure 6.1), it appears that particular concentrations of productive activity are emerging in the new towns of Livingston, Glenrothes and East Kilbride, rather than in the older urban or industrial areas. Of particular note, in addition, is the concentration of independent design houses in the Edinburgh area. All of these are spin-offs from the electronics department of Edinburgh University.

With Scotland beginning to emerge as probably the most important semiconductor production complex in Europe, the savings that can be made on transactional costs, given continuing supplies of scientific/technical and unskilled labour, may represent compelling reasons why the bulk of inward semiconductor investment in Europe may seek to locate there rather than at other possible sites. The significance of this observation for economic and social development in Scotland will be taken up in the next chapter. For now, however, we need to address the relationship of semiconductor production there to the international division of labour in which it is embedded.

Semiconductors, Scotland, and the international division of labour

I began this study by arguing that the significance of industrial restructuring in any given territorial unit could only be fully grasped if it was understood not only within that unit itself, but also, and crucially, in relation to the structure and dynamics of the international division of labour within which it 'lives' or 'dies'. This section, then, explores some of the 'international issues' associated with semiconductor production in Scotland.

Figure 4.2 (p. 56) summarises the international dispersal of the labour processes of those American companies who in 1985–86 manufactured semiconductors in Scotland. Before discussing the direct connections between semiconductor labour processes in Scotland and those in other offshore locations, it might be helpful to reiterate here some of the general features of the industry's dispersal around the globe.

As we might expect, the majority of the labour processes together with the bulk of the managerial functions are, for the companies with which we are concerned, concentrated in the United States. Five points, however, need to be emphasized:

1. Research and Development (R & D) takes place only in the USA. Although some circuit design centres have been set up in a

Figure 6.1 **The Scottish semiconductor production complex**

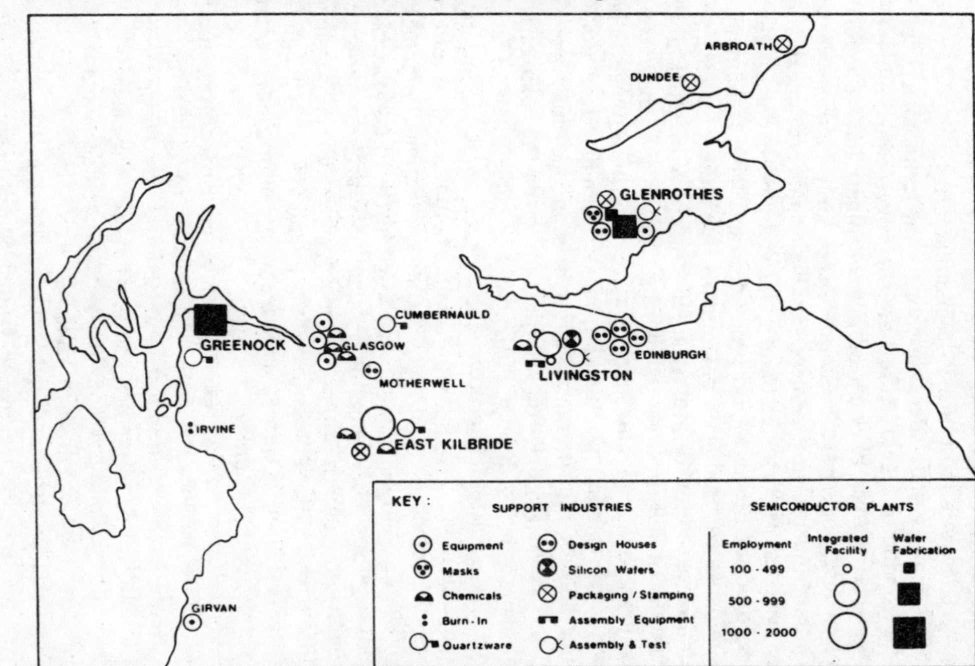

Sources: Interviews with company executives; company data; locate in Scotland (1984a, b); Baggott (1985b).
Note: For technical reasons the map does not depicit all support industries. It under-represents, therefore, the 'density' of the production complex.

number of other industrial societies, principally in Europe, but now increasingly in selected East Asian locations, they are not engaged in innovative research. Their function is to adapt basic designs to specific customer requirements. Even recognising that design centres do not have an innovative role, it is significant that in the Scottish case there is neither R & D nor design centres associated with most of the US manufacturers.

2. Mask making and wafer fabrication – the technological cores of semiconductor manufacture – are also concentrated in the USA. While a number of wafer-fabrication plants have been set up elsewhere, they have been established only in those societies capable of supplying the necessary high-quality technical labour, as well as the forms of managerial and environmental control essential for the high-yield production of such a technologically sophisticated product. Consequently, with the very recent and limited exception of Malaysia, US firms have invested in wafer fabrication only in advanced industrial societies. In the European context, Scotland has been of particular importance. Unlike wafer fabrication, mask-making has not been dispersed at all.
3. The bulk of the routine assembly of semiconductors has been dispersed to Third World societies, and given the companies we are concerned with here, particularly to some of those in East Asia. While assembly processes involve low capital-labour ratios and require large supplies of unskilled labour, some automation has taken place in recent years. Significantly, some companies are now developing fully automated assembly facilities in advanced industrial societies, including Scotland. Motorola in particularly significant in this context. Indeed, one of the significant differences between the entries for Scotland in Figures 3.2 and 4.2 is the evidence of recent investment in (automated) assembly facilities in that country.
4. Final testing is now a relatively capital-intensive process and has tended to be located in the manufacturing plants that have emerged in advanced industrial societies. As we saw in Chapters four and five, however, some testing facilities have been set up in the more 'advanced' Third World locations such as Hong Kong and Singapore as well as in Malaysia.
5. Corporate control, and hence key decision-making processes remain firmly rooted in the United States. The only partial exception is with Motorola which tends to give more local autonomy to its plants (or regional autonomy in the case of its East Asian operation centred on Hong Kong) than is normally the case.

The implications of these general features for semiconductor production in Scotland and East Asia will be analysed in the next chapter. Here, however, we need to chart the connections between labour processes in Scotland and those in other offshore locations of the respective corporations.

The American-owned semiconductor industry in Scotland does not have a full production capacity in the sense that it is substantially based on the operation of only one labour process: wafer fabrication. It is, therefore, heavily dependent on R & D and mask production which takes place in the United States, and to a lesser extent on design facilities which tend to be located in England, or elsewhere in Europe, but not, as I have already indicated, in Scotland itself. Much of the scientific and technical knowledge (particularly its innovatory forms) on which the industry in Scotland depends, therefore, is not under the control of the Scottish plants, nor is it generated in Scotland (or Europe for that matter). Only National Semiconductor and Burr-Brown currently operate design centres in the country (but their design teams concentrate on adapting for European customers circuit designs which originate in the United States) and Hughes use a Scottish subcontractor to manufacture masks that were designed in their US facilities.

As with wafers fabricated in the USA, Scottish-produced wafers are assembled and tested in East Asia and then dispatched directly to the customer. National Semiconductor assembles and tests its Scottish-produced wafers in its Malaysian and Thai plants. General Instrument does the same in its Taiwanese plant, and Hughes assembles under a subcontract arrangement (with Semiconductor Devices; see Chapters 4 and 5) in the Philippines and Hong Kong, but tests the final product after return to its Scottish plant. So far, only Motorola has developed a different arrangement. Though previously assembling its semiconductors in East Asia, Motorola now has a fully automated assembly and test facility alongside its wafer fabrication plant. The spatial combination of these labour processes in Motorola's Scottish plant constitute it as the most important production facility in the company's operation outside of the United States, and probably the most important semiconductor plant in Europe.

When we examine American semiconductor production in Scotland in the light of the industry's overall international division of labour, it is clear that the country is in a far stronger position than any of the territorial units in the East Asian regional division of labour. This comment is as true of its ability to penetrate, without tariff restrictions, the world's third largest semiconductor market (Table 6.6), and hence avoid a possible future intensification of protectionist measures (see Chapter 7) as it is of the nature of its labour processes.

Although test facilities have been installed by some US plants in certain East Asian locations, and some automation of assembly processes has taken place, (for instance in National Semiconductor's Malaysian plants), it remains the case that the technological core of the industry (mask production, wafer fabrication) does not exist in any American plant in that region with the exception noted above. However, it is also clear that the Scottish industry has major weaknesses. It does not have a capability for producing (i.e. fabricating) the most technologically advanced semiconductors, such as microprocessors, nor, in particular, does it have a fully integrated production capacity. Consequently, it is in an exposed position insofar as it remains dependent on foreign technical expertise and subject to overseas corporate control.

In this latter sense, the Scottish industry is subject to decision-making processes that are spatially far removed from the possibility that workers or local governments may be able to exert pressure upon them.

Furthermore, while the labour processes in which American firms have invested in Scotland help to ensure that it is unlikely that a significant reduction in the scale of their operation will occur (unlike, for instance, in Malaysia during the recession of the mid-1980s: McGee 1985; Salih and Young 1987), it remains the case that 'Silicon Glen' is not yet a self-sustaining semiconductor production complex. While this situation persists, and in the context of the current global economic crisis, American (or foreign) dominance over semiconductor production in Scotland remains a feature of some concern. This having been said, however, the point needs to be made that in the context of global restructuring within the industry and the transnational financial and technological alliances associated with it (e.g. between Motorola and Toshiba) it is unlikely that *any* country or region will again be capable of developing a fully autonomous semiconductor industry (Ernst 1987). This point, together with a number of other issues surrounding the continued development of the industry in Scotland, as well as in East Asia, will be explored more fully, however, in the next chapter.

Chapter seven

Prospects for globalisation and development

In this study I have analysed some of the social and spatial processes associated with the globalisation of the American semiconductor industry. Distinguishing between the internally necessary and historically and spatially contingent determinants of the industry's growth and evolution, I have shown how they articulated to produce the world's first semiconductor production complex, Silicon Valley, California. The continued development of those determinants, however, generated contradictions in Silicon Valley and elsewhere in the United States. Growing Japanese competition in transistor markets ensured that the industry could no longer survive without significantly restructuring its organisational form. Given that its labour processes were technically disarticulated, the US semiconductor industry was able to restructure by means of taking the global option. In the context of favourable US import regulations, its labour-intensive, low value-added labour processes were dispersed to selected Third World sites, particularly in East Asia, in order to take advantage of their supplies of cheap labour. Later, certain high value-added and capital-intensive labour processes were internationalised partly in order to penetrate major, but protected, markets. The emergence of US semiconductor production in Europe was a case in point. By the mid-1970s, semiconductor production had become a paradigmatic example of an industrial branch managerially, technically, and spatially organised broadly according to the principles identified by the 'new' international divison of labour thesis.

Within a few more years, however, the contours of the industry's international division of labour, had begun to change drastically. This was especially the case in the developing countries of East Asia, where, partly as a result of state intervention, partly because of the availability and cost of highly skilled engineering and technical labour, certain territorial units began to emerge as the technological and managerial cores of a specifically regional division of labour. As a result, these units, as part of their more general transformation from peripheral to semiperipheral status within the world-system, were able to loosen their structures of dependency internationally, while further subordinating the peripheries of the regional division of labour.

This changing international division of labour, coupled as it has been with a new mode of industrialisation, as well as processes specific to particular territorial units themselves, has led to a variety of development possibilities. In some cases, as with the gang of four, these possibilities confirm that under certain circumstances, capitalist industrialisation can lead to genuine development in the Third World. Rather than summarize the arguments of this study in any greater detail, however, I wish to outline some of these development possibilities for the territorial units with which this book, primarily, has been concerned. In the final chapter, we will assess the significance of our work for the theorisation of the international division of labour.

Global restructuring

The prospects for all territorial units, including those regions of the United States where the semiconductor industry's principal production complexes are located, are clearly determined in part by the changes the international divison of labour is currently undergoing and is likely to continue to undergo for the foreseeable future. We begin, therefore, by indicating some of the key structural shifts to which the industry is currently subject at the global level.

The form of the international division of labour discussed by the NIDL thesis was the spatial outcome of an earlier, albeit quite recent, round of restructuring. Restructuring in the late 1980s and early 1990s would seem to imply quite different spatial arrangements, at least as far as the industrialisation (or reindustrialisation) of those territorial units discussed here is concerned.

Specifically, there seem to be certain processes at work which in combination tend towards both the spatial reintegration of production sites with the principal markets of the developed world, and the vertical integration of some (though not all) labour processes in selected production sites within the developing world. As a corollary to the latter tendency, particular territorial units are beginning to specialize in advanced design work and in the managerial control of region-wide production operations.

Let me begin by outlining some of the processes which compose the first tendency I have identified. Here we need to bear in mind that, in spite of an expansion of the East Asian and other developing country markets for semiconductors and other commodities in recent years, the principal markets for manufactured commodities (including semiconductors) remain the USA, the EEC, and Japan. Indeed, these markets have remained remarkably stable, to the tune of 70 per cent of world markets throughout the 1970s and early 1980s (Ohmae 1985). The particular forms the processes associated with the reintegration of semiconductor production sites and principal markets seem to be taking, are as follows:

1. In the context of a collapse in demand for standardised semiconductors (associated with the saturation of consumer electronics markets) and hence a deepening crisis affecting both major US merchant producers and their Japanese counterparts, 'strategic alliances' are beginning to emerge between US and Japanese firms (e.g. Motorola and Toshiba; AMD and Sony; RCA and Sharp; Intel and Oki; Zilog and Hitachi; Fairchild and Hitachi), US and European firms (e.g. Motorola and Thomson; Texas Instruments and Philips; National Semiconductor and Thomson; Silicon Systems and Ferranti), Japanese and European firms (e.g. Toshiba and SGS; Toshiba and Siemens), European firms themselves (e.g. Siemens and Philips) as well as between US companies (e.g. Motorola and AMD; Motorola and Signetics; National Semiconductor and Chips and Technologies). These alliances are designed to jointly finance and exploit increasingly expensive technological innovation (particularly with regard to new-generation ICs and customised devices) and hence corner important segments of the market, which itself is increasingly dependent on innovation for its profitability. Partly as a result of the enormous capital sums now required to support innovation, US merchant producers additionally are beginning to be absorbed by US and other electronics system houses. Thus, for instance, Signetics, which was absorbed by the Dutch electronics giant, Philips, in the 1970s, now also has technological links with the US computer firm, Honeywell, as well as with Motorola; and Intel is linked with NCR to develop advanced customised devices.[1] One should note here, that it is the US merchant houses on which Asian-Pacific production has depended, and that the captive divisions of US system houses have hardly ever assembled their semiconductors abroad, but rather have managed to maintain cost-effectiveness by automatically assembling their product in the USA.
2. While the economics of offshore assembly have not deteriorated to the extent that there has been the switch to onshore, automated assembly predicted by some commentators (for instance, Rada 1982), there are developments in both the

technology of automated assembly and in the demand for particular types of semiconductor that in due course may have significant implications for the industry's spatial division of labour, and (though in different ways) for many of the locations within it. Amongst the reasons for this are that:

(a) '(A)ssembly technology is becoming ... extremely sophisticated. Hourly wages are no longer the key issue. The constant updating of equipment and the need to use frontier technology is now the overriding factor.'[2]
Under these circumstances, territorial units at the periphery of the East Asian division of labour, such as the Philippines, Thailand, Indonesia, and to a lesser extent Malaysia, at first sight would seem to be exposed, as their supplies of cheap labour usually have been taken to be their principal productive advantage. However, this may not necessarily be the case, as we shall see below.

(b) There is an increasing demand for customised semiconductors (see Table 7.1 for current markets). In part, this is a product of the increasing significance of 'defence' markets as the United States and its allies gear up for war, but also of rising demand for customised products in consumer electronics markets. With the rise in customised demands has come an increasing concern with product quality. Major consumer electronics customers are now beginning to join their military counterparts in insisting on the involvement of their engineers in product assessment at each stage of the production process. Clearly where the labour processes of

Table 7.1 World market for standard and custom integrated circuits, 1986

National Market	*Standard* *$m*	*Custom* *$m*	*Total* *$m*
US	6,490	1,095	8,429
Japan	7,586	866	9,852
West Germany	848	142	1,102
Britain	769	148	989
France	526	18	684
Italy	300	42	390
Totals	16,519	2,374	21,446

Sources: Dataquest; Mackintosh, Butler & Cox: Compiled by *Financial Times* (31 March, 1987)

the industry are spread across the globe, then customer monitoring becomes more expensive and less efficient than would be the case if the labour processes were spatially concentrated.

(c) Given the increasingly sophisticated nature of semiconductor technology, there are now considerable benefits to be gained from assembling the devices automatically. In particular, automated assembly has the capacity to ensure a higher quality and yield. This is the case especially for the more advanced semiconductors based on VLSI (very large scale integration) technology, and thus requiring the 'bonding' of many more connections than simpler devices. Furthermore, continuing innovation in the assembly technology itself, has resulted in machines that can be easily re-programmed. As a result, automatic assembly can now be used for the small-batch production runs associated with customised semiconductors. Both Fairchild and National Semiconductor have developed fully automated assembly lines at a number of their US plants, as have Motorola at its plants in Arizona and Scotland. As a consequence of automation, Motorola's Scottish plant, for instance, no longer assembles wafers in East Asia.

3. While most of the restructuring that the semiconductor industry has undergone in the forty years of its existence, has been technological and spatial, there are now indications that the labour processes themselves are about to be re-organised. The spectacular commercial success of Japanese industrial corporations has led to a number of Western companies experimenting with the *kanban* or 'just-in-time' manufacturing system. This system requires highly synchronised and continual deliveries of components and materials from the company's suppliers. As a result, it is capable of significantly reducing storage and overheads costs. It can also lead to a reduction in the size of the labour force that would otherwise be necessary, and it helps improve productivity (Sayer 1986b). In the automobiles sector, General Motors, for instance, is developing the kanban system as part of its Saturn project in the USA (Hill 1987), and especially significant for our purposes is that one of the world's leading semiconductor producers, National Semiconductor, is experimenting with the system in its US plants (Hong Kong Productivity Centre, *Electronics Bulletin*, November, 1985: 24).

Whether or not the National Semiconductor experiments are successful, the kanban system seems destined to do for labour processes and productivity in the 1990s what Taylorism and

Fordism did for them in an earlier phase of industrial capitalism. As a result, kanban could become a major basis on which industries, including semiconductors, are restructured in the coming years. A key element of the kanban system is that it requires that suppliers produce in spatial proximity to their principal customers. In other words, it is a manufacturing system that would help to compound the other tendencies towards spatial integration mentioned above.

4. In addition to the growing significance of markets for customised products and the organisational and technical developments indicated above, many parts of the United States where semiconductor – and other electronics producers – are located, have been recipients of low-wage, migrant labour particularly from Latin America, the Caribbean, and Asia. As a result, labour processes which, perhaps only ten years ago, were uneconomic in the American context (including electronics assembly), have been resurrected in a number of US cities. As Sassen-Koob (1987) has argued, the restructuring which the US economy has undergone over the last ten to fifteen years, has created the conditions for new rounds of accumulation inside the American urban heartlands. One of these conditions has been the emergence (or deliberate creation) of new supplies of unorganised, relatively docile, cheap labour. The consequence of this situation, she suggests, is that it tends to encourage the reconcentration of industrial capital in the core economies.
5. The deepening global economic crisis has led to increasing demands for protected markets in the core economies. In addition to the strategic alliances indicated above, US, Japanese, and European companies are rushing to set up plants inside each other's boundaries. In part this is a response to protectionist pressures. Thus in addition to the US companies with manufacturing plants in Japan (see Figures 3.2 and 4.2), major Japanese semiconductor producers such as NEC, Hitachi, Toshiba, Fujitsu, and Mitsubishi all have wafer-fabrication and assembly plants (with automated lines) producing memory ICs in the United States (Hong Kong Productivity Centre, *Electronics Bulletin*, November 1983: 10). As we saw in the previous chapter, NEC also has a fully integrated plant in Scotland, and other Japanese producers setting up plants in Europe can be expected in the near future. The economic unification of the EEC in 1992, rendering it the world's largest single market, will undoubtedly compound these tendencies.

In spite of the processes indentified above, there is, as yet, hardly any evidence for the semiconductor industry, that disinvestment on the East Asian periphery or semiperiphery has resulted in the relocation of productive capacity to the developed economies. What disinvestment has taken place, such as by Fairchild and Sanyo in Hong Kong and Fairchild in Indonesia, has, if anything, resulted in a redistribution of productive capacity within the developing East Asian region itself.

What seems to be of greater significance is that in certain locations within the East Asian region, companies seem to be interested in capital deepening together with the vertical integration of particular labour processes. As I have argued in previous chapters, capital deepening has been particularly pronounced among US semiconductor houses in Hong Kong, Singapore, and the gang of four generally. What we have seen most recently, however, is the willingness of at least one major US manufacturer to invest in automated assembly, testing, and wafer fabrication in a vertically integrated operation in Malaysia. While the specific reasons for this development were indicated in Chapter four, the general reasons are probably that, as Dieter Ernst has suggested:

> Any change in the current pattern of manufacturing and sourcing would involve substantial existing costs, both in terms of closing down plants, reshuffling supply and market networks, and in terms of benefits foregone that could be reaped from achieving even higher stages of internationalisation.
>
> (Ernst 1985: 343)

It is important, however, to recognise that if the incidence of integrated plants within the East Asian regional division of labour increases in the next few years, they are likely to be concentrated in those locations which, in addition to other production factors (necessary and contingent) are capable of supplying in reliable quantities, technical and engineering labour power. While Malaysia, seemingly, must now be taken as a case in point, this would appear to suggest, predominantly, the gang of four. It is necessary to bear in mind, however, that capital deepening and vertically integrated production in the developing societies of East Asia has so far only been associated with the technologically simpler, standardised semiconductors rather than with more advanced products such as microprocessors or customised devices. Given this divergence in the technological nature of the product generated from diverse locations within the overall international division of labour, the tendencies of spatial integration in the core economies (producing the more technologically advanced semiconductors) and the emergence of vertically integrated plants in the periphery or semiperiphery (producing less-advanced products), can be seen as complimentary rather than contradictory. These and other issues will be addressed in more detail, however, when we deal with the implications of these tendencies for the development prospects of the territorial units on which our discussion has concentrated.

Territorial implications

If the tendencies identified above continue to emerge and thus affect the global restructuring of the semiconductor industry, what are likely to be the implications of this for the industry's future development in East Asia and Scotland?

East Asia

The first point that must be made is that there is unlikely to be a wholesale exodus of US and other foreign semiconductor investment from East Asia. On the contrary, as I showed in Chapters four and five, investment has increased in recent years (see Table 5.7 on the Hong Kong case, for instance), including in automated assembly processes, the very technological innovation that eliminates the need for large supplies of cheap unskilled labour, and therefore supposedly (according to Rada 1982, for instance) the productive advantage of territorial units such as those of East Asia.

While US semiconductor investment is unlikely to withdraw from anywhere in the region (with the exception of its very recent withdrawal from Indonesia noted above), it is likely to concentrate increasingly in those territorial units that I have referred to as the cores of the regional division of labour. Like any spatial division of labour in the context of capitalism, the East Asian division has developed unevenly. With the emergence of semiconductor and more general electronics production complexes in the regional cores, as I argued in Chapters four and five, it is likely that development processes in those cores have now taken on a logic of their own. As a result, new semiconductor investment, particularly in more advanced labour processes, is likely to be attracted to those locations rather than others in the region. Consequently, the peripheries of the regional division of labour (the Philippines, Thailand as well as Malaysia) are likely to continue to concentrate on the assembly (and in some cases the testing) of standardised, technologically simple (in relative terms) semiconductors. Some automation of the assembly process has already taken place (in Malaysia, for instance), and this is likely to continue, with consequent negative implications for unskilled employment opportunities. The presence of foreign-owned semiconductor manufacture in the regional periphery is unlikely now to stimulate indigenous developments, perhaps irrespective of the development strategies that the various national governments may wish to pursue. Other than providing employment (in declining numbers) and the economic effects of paying wages, utilities and taxation bills, the foreign semiconductor presence in these societies is likely to remain parasitic on the local economies. Their presence there, then, is likely to continue to constitute what I previously referred to (in Chapter 2) as development *in* but not *of* those territorial units.

As I have indicated already, the possible exception to this scenario for the peripheries of the regional division of labour, is Malaysia. The signs are that investment in wafer fabrication facilities there, by a number of US and Japanese companies, is likely to continue. While Malaysia, then, may be on line to emerge as the foreign-owned wafer fabrication centre of the developing world, at this stage it should not be reconceptualised as one of the cores of the regional division of labour. The reason for this is not simply the absence of a locally owned production complex (this barely exists in Singapore either). Rather it is the logical outcome of two other factors previously mentioned. First, that region-wide managerial control, to which the Malaysian plants are subordinate, is based not in Malaysia, but typically in Singapore or Hong Kong; and second, that crucially important circuit-design work is often associated with those regional control centres, if they exist in the region at all. Thus despite the indications that vertically integrated plants founded on technologically advanced labour processes are beginning to emerge in Malaysia, this does not necessarily imply, in any simple way, that the entirely foreign-owned semiconductor industry there will be more autonomous and less dependent than in the past. Nor does it suggest that the industry necessarily will become any more integrated and less parasitic on the local and national economies than has been the case in other units on the periphery of the regional division of labour.

Semiconductor production in the cores of the regional division of labour displays different, but not opposite, developmental possibilities to those outlined for the peripheries. While, as I have suggested above, these territorial units are likely to be the recipients of increasing investment in more advanced labour processes, including wafer fabrication, and possibly even mask making, they are unlikely to become centres of innovative R & D. They will continue, in other words, to be locked into and constrained by the organisational and technological dependencies associated with their semiperipheral status in the overall international division of labour.

This general comment, however, cloaks significant differences which arise from within the particular territorial units themselves. Of special note are the situations of Singapore and Hong Kong. As I indicated in Chapter four, Singapore, alone of the gang of four, has failed thus far to develop a significant indigenously owned presence in semiconductor manufacture or electronics more generally. As a result, the production of over 70 per cent of the country's manufactured exports are controlled by foreign corporations as is over 80 per cent of electronics production (Mirza 1986: 5, 104). While one Singaporean scholar has recently argued that this extreme dependency on foreign managerial decisions, finance, and technological know-how does not constitute a problem for the country's future development (Lim 1987) it seems to me that in the context of some of the structural tendencies indicated above, Singapore's economy, based as it is predominantly on semiconductors and electronics, is badly exposed to the vagaries of the world economy and geo-political considerations.[3]

As I mentioned in Chapter four, however, the Singaporean government, if anything, seems to be attempting to by-pass the manufacturing stage of industrialisation and move directly to a situation where economic growth, in part, becomes based on its ability to sell its technical knowledge and expertise (in the form of various types of software and design work). The emergence of integrated production facilities in Malaysia, given that a number of the companies manufacturing there already have their regional headquarters and design centres in Singapore, could well enhance these efforts to turn Singapore into one of the 'knowledge centres' of the region, as far as the semiconductor and electronics industries are concerned. Whether this strategy, in terms of economic growth, will be successful is one thing, but what it will mean for relative inequality in the country is another. Advanced software production etc. does not employ many people, and the service jobs which it could generate are not known for the high salaries they pay. Without mechanisms to redistribute the wealth so generated, the seeds of future social conflict could well be sown by this particular state development strategy.

The problems which confront semiconductor production in Hong Kong and its capacity to continue to assist the territory's economic development, are of a rather different order. First, semiconductor, and indeed all forms of manufacture, are now being affected by the gradual absorption of Hong Kong into the Chinese economy. A consequence of this situation is that those labour processes that still rely on cheap unskilled labour are being increasingly relocated to the Chinese mainland, in search of yet more opportunities for absolute surplus-value creation. Already some 10,000 Hong Kong-owned factories, reputedly employing in excess of one million people, have been set up in Guandong Province (immediately adjacent to the Colony) alone, predominantly producing cheap clothing and electronics.[4] Of the foreign semiconductor firms in the territory, National Semiconductor plan to assemble in China in the near future, while Sanyo have recently (1987) fled Hong Kong completely, transferring production to a new plant in the Shenzhen Special Economic Zone, just north of the border. While locally owned electronics firms could well continue to shift production to China, at least for their lower-grade products, it is unlikely that foreign or local semiconductor producers are likely to follow Sanyo's example. What is much more likely, is that Hong Kong will become the technological core and managerial-control centre of a domestic Chinese division of labour in semiconductor production. Design work, wafer fabrication (by Hong Kong and foreign firms) and testing will probably remain in Hong Kong while labour-intensive assembly processes will be increasingly relocated to China. Such developments, however, will compound the declining contribution which manufacturing industry makes to employment in the territory (see Chapter 5), which in the context of the deepening world economic crisis and Hong Kong's extreme exposure to the vagaries of the world-system (by virtue of the absence of a domestic market), could have serious social and political consequences prior to the territory's formal absorption by China in 1997.[5]

While the prospects for a continued major presence of US semiconductor firms in the developing countries of East Asia vary depending on which territorial unit we are talking about, what seem to be the prospects for those indigenous firms that have emerged in the region?

First, as I mentioned in Chapter four, we must take cognisance of the fact that these indigenous producers manufacture predominantly low-grade (in technological terms) discretes and memory integrated circuits, of the type used in televisions, VCRs, and simpler microcomputers and watches. The expertise, technology, and quality-control conditions necessary for the production of higher-grade memory ICs, and certainly for the most technologically advanced semiconductors – the microprocessors – remain firmly locked in Japan and Europe (memories) and the USA (microprocessors and memories) (Ernst 1987; Johnstone 1988). Furthermore, the indigenous firms are dependent on Japan and the United States both for R & D, circuit designs, and for the manufacture of celluloid circuit masks.

If these firms are content, however, to produce for their domestic, and other Third World markets, then these constraints may not be particularly burdensome. However, under these circumstances, their growth and hence their contribution to their respective local economies, will be severely restricted by the small size of all semiconductor markets other than the US, Japan, and the EEC. If they attempt to break into the major markets, however, they confront all of the problems outlined above. In addition, however, they confront the problem that success in the major consumer electronics markets is now becoming less associated with price competition (producing more cheaply the same product as one's competitor) and more with product competition. Increasingly, then, a premium is beginning to be placed on the technological superiority of one's product. Gaining the technological edge over one's competitors, requires not only availability of high-quality scientific and engineering talent, but also an educational system capable of reproducing that talent from generation to generation. In addition, it involves the investment of increasingly large sums in innovative R & D and state-of-the-art production technology. While the South Korean and Taiwanese governments are making considerable efforts in this direction, the Hong Kong government is not, thus putting at further risk the territory's electronics industry.[6] In spite of the efforts of certain states to develop an R & D capability, however, when we take into account that their national producers are confronted with the strategic technological and financial alliances between market leaders and between major merchant producers and giant systems corporations indicated above, one must wonder how any of the firms based in the gang of four can in the future hope to compete effectively in world markets.

That said, it is undoubtedly the case that Taiwan and South Korea will continue with their latecomer industrialisation strategies, based as they are, in part, on electronics and semiconductor production. I have already mentioned (in Chapter 4) that South Korean firms in particular, have attempted to tap American technological expertise by setting up R & D facilities in Silicon Valley. In addition, a number of US firms have developed production arrangements with South Korean manufacturers. This is particularly the case with small, Silicon Valley, customised producers, who, rather than invest in their own basic wafer fabrication capacity, utilise that of Korean producers. Thus firms such as LSI Logic, and Chips and Technologies, use Hyundai and Samsung respectively to fabricate all but the last two circuit layers. These last two layers, which carry their own proprietory designs, are fabricated in their own facilities in Silicon Valley. In this way the US producers retain control over the technological core of the devices, and thus little, if any, state-of-the-art technology is transferred to the Korean producers.

However, if any East Asian semiconductor industry is capable of becoming a major actor in world markets, then the South Korean industry has by far the best chance. By means of the financial and technological alliances between private capital and the state indicated in Chapter four, South Korean manufacturers are attempting to travel the Japanese route to 'big-league' status. They already have a protected, growing, and potentially large domestic market[7] and are beginning to develop potentially important technology transfer agreements, of which the latest (1986) has been between Hyundai and General Instrument (*San Francisco Examiner*, 9 November 1986).

Scotland

If the prospects for semiconductor production in East Asia are variable, depending on the territorial units in question, what of the situation in Scotland?

There is unquestionably already a strong, and expanding semiconductor industry in Scotland. Much of its strength lies in the fact that it specialises in one of the most technologically advanced labour processes, wafer fabrication, and produces wafers for both the European (currently a buoyant semiconductor market) and world markets. Its presence has encouraged investment by foreign supplier firms and it has stimulated the emergence of locally owned semiconductor manufacturers, suppliers, design houses etc. Inspite of these encouraging developments, however, a number of problematic issues remain.

1. The industry's strength in wafer fabrication is also, paradoxically, a sign of its weakness. The continued improvement of semiconductor technology and hence the sophistication of the wafers depends upon basic research, circuit design, and mask manufacture. Only two of the American corporations have design centres as part of their Scottish operations, and none of them engage in basic research or mask production. Thus, while the continued development of the US section of the industry may stimulate the demand for highly skilled scientific and technical labour, that labour, in the event, is not being applied to the technological innovation so essential to the industry's growth, and indeed survival (Cooke, *et al.* 1984).

   Fortunately, the prospects in the indigenous electronics sector, are in this sense, more positive. These firms are undertaking basic R & D not only to a greater extent than their multinational counterparts in Scotland, but also, seemingly, to a greater extent than similar small high-tech firms in Silicon Valley (Oakey 1984). In addition, it is the small indigenous firms, rather than the multinationals that maintain close research links with the universities. Indeed, a number of the independent design and software houses in Scotland are 'spin-offs' from Edinburgh University (e.g. Lattice Logic, Wolfson Microelectronics Institute, Walmsley Electronics).

2. While the emergence of supplier manufacturing plants in Scotland may alter the situation, the research on American semiconductor plants reported here confirms the view that they, like other foreign-owned plants, have relatively few linkages to the local Scottish economy (cf. Young 1984). Only about 10 per cent of Motorola's annual expenditure, for instance, goes into the Scottish economy (and then mainly for cleaning, catering, maintenance services, and utilities) whereas the equivalents for National Semiconductor and General Instrument are 30 per cent and 5 per cent respectively. As the first two figures include utilities costs, the only additional ways in which the plants would appear to stimulate the Scottish economy is by their payment of property taxes (corporations tax being a central government levy) and via the 'knock-on' effect of their wage payments. Rather like their counterparts in East Asia, then, American semiconductor plants in Scotland would appear to be relatively parasitic on the local economy.
3. Though the investment by such companies as Motorola, Indy Electronics, and NEC in automated assembly and test equipment may be seen as a positive development for semiconductors in Scotland (in the sense that it advances the tendency to spatially re-integrate semiconductor labour processes and hence strengthen productive capacities there) the potential benefits are not clear-cut. Though the automation of labour processes tends to result in increased demands for technical labour power, it does little to expand job opportunities for semi- and unskilled workers. Thus, this form of industrial restructuring helps both to compound the processes of class restructuring as evident in Scotland as elsewhere in Britain in recent years (Massey 1983), and additionally it does little to improve the material circumstances of those who constitute the bulk of the Scottish unemployed.
4. Finally, semiconductor production in Scotland, together with many of its support industries, is, as in East Asia, substantially controlled by foreign corporations. As a result, it is subject to decision-making processes that are spatially far removed from the possibility that workers or local governments may be able to exert pressure upon them. While the labour processes in which American firms have invested in Scotland help to ensure that it is

> unlikely that a significant reduction in the scale of their operations will occur, it remains the case that 'Silicon Glen' is not yet a self-sustaining semiconductor production complex. While this situation persists, and in the context of the current global economic crisis, American (or foreign) dominance over semiconductor production in Scotland remains a feature of some concern.

The thrust of this chapter has been with the implications of certain restructuring tendencies at the global level for the continued development of essentially 'national' semiconductor industries. Before we return to more theoretical issues in the final chapter, a point made earlier needs to be recovered and emphasized. As I have indicated, one of the forms the global restructuring of the industry is taking is the emergence of financial and technological alliances between companies which span national boundaries. If these alliances continue to develop such that entire 'national' companies (not merely plants within companies) begin to specialise in the production of particular parts of the semiconductor and/or in particular forms of semiconductor technology, then the conditions for a truly 'multinational' semiconductor industry may have been achieved. Under such circumstances even the most advanced semiconductor industries, such as those of the United States and Japan, are likely to become locked into structures of dependent development. While the structures of dependence will develop asymmetrically such that certain organisational and technical components of the multinational production system will have more 'relative' autonomy than others, it may soon become unlikely that any country will again be capable of developing a fully autonomous semiconductor industry (cf. Ernst 1987).

Chapter eight

Semiconductors, development, and the changing international division of labour

I began this study by suggesting that the dynamics and implications of industrial restructuring need to be grasped not only in relation to the particular territorial unit, but also in relation to the way in which the industrial branch, or complex of activities in a given location, articulates with those in other locations and with the structure of the world economy as a whole. I suggested that a potentially fruitful way of assessing the local and international dimensions, at one and the same time, was by a critical application of the NIDL thesis. In this concluding chapter, I wish to reverse the process and indicate some of the implications of our discussion of semiconductor production in East Asia and Scotland for the theorisation of the international division of labour. We begin by indicating some of the general implications of our work for the territorial units with which we predominantly have been concerned. We then turn to examine specifically the NIDL thesis in the light of the foregoing analysis and finally we look once more at semiconductors, electronics, and similar forms of high-technology production as a new mode of industrialisation.

At the most general theoretical level, what is clear from our analysis of the globalisation of semiconductor production and its developmental consequences, as well as from other research summarized by Castells (1986b), is that it is no longer possible to understand local economic and social change in a global context by means of the core-periphery categories. In spite of my use of them for heuristic purposes, they appear to be simply too static and inflexible to be able to grasp the enormous range of development actualities and possibilities that exist within industrialisation (and other economic development) processes in various parts of the globe. Though the notion of semiperipheral territorial units operating as 'way-stations' for economic and productive transactions within the international division of labour, may go some way in helping to resolve our difficulties (as I argued in Chapter two), it still does not seem to take us far enough. Scotland, for instance, may be developing a strong base in semiconductor production, but it still lacks a fully integrated production system and is massively dependent on corporate decisions and technology that originate in the United States. Though the emergence of a local semiconductor production complex is encouraging, Scotland is still a long way from having a viable, self-sustaining semiconductor industry. It is in this sense that reference to the central belt of Scotland as 'Silicon Glen' (thus evoking images of the world's most dynamic and successful high technology complex, 'Silicon Valley', California) is both misleading, and analytically, potentially dangerous. By the same token, Scotland cannot be regarded as a 'semiperipheral' location for semiconductor production. Though we might seek to regard it as such for heuristic purposes, when we compare its performance with other territorial units potentially making claims to be included in this category – specifically the East Asian gang of four – Scotland clearly has a far stronger, technologically and managerially more autonomous role in the international division of labour, than any of them, with the possible exception of South Korea.

As for the East Asian territorial units themselves, they exhibit considerable diversity, associated partly with their role in the international division of labour, but partly, and crucially with internal historical developments – including varieties of state intervention – specific to the particular units themselves. As a result of the articulation of these elements, some 'semiperipheral' units in the region – namely the gang of four – have development possibilities that are unlikely to be matched in the foreseeable future by the other units. Even within the gang of four, themselves, however, there are key structural and geo-political differences that raise major questions, for instance, about the future viability of Hong Kong as an industrialising society.

In essence, what we are confronted with is a changing international division of labour which is becoming less organised on the basis of cores, peripheries, and semiperipheries. Certainly the international division of labour is determined and driven by its relations of dependency. Thus it still makes considerable sense to speak in terms of processes of 'dependent development' controlled from 'cores' and 'world cities' (Friedmann 1986). But the problems of identifying the peripheries and semiperipheries seem to be becoming ever more difficult. For reasons I shall advance later, however, this increasing flexibility in the world-system, unfortunately does not mean that we can now look forward to the 'end of the third world' (cf. Harris 1987), at least as a material reality.

As regards semiconductor production, a continuum of interrelated production activities seems to have emerged at the global level. At the top end, it stretches from the United States and Japan with their relatively autonomous, but increasingly interconnected (by virtue of the 'strategic alliances') semiconductor industries, through the emergent complexes in Scotland, West Germany, and other EEC countries where high grade semiconductors are also produced, but in a context of dependency on the USA and increasingly Japan. Lower down the scale we have the semiconductor industries of South Korea, Taiwan, Hong Kong, and Singapore. While these locations play a more subservient role in the international division of labour than does Scotland, with the exception of Singapore, they do have indigenously owned production facilities. These are, however, at present capable of producing only relatively low-grade semiconductors. Though the role of this latter group of countries in the overall international division of labour may be relatively limited, they themselves have emerged as the 'cores' of an East Asian regional division of labour. Finally, on the fringes of the regional and international divisions of labour are countries such as the Philippines, Thailand, and Indonesia which continue to be involved largely in the assembly of standardised, relatively less technologically sophisticated semiconductors. Given their highly restricted position within the global arrangement of the industry, their chances of developing their own self-sustaining semiconductor industries must be regarded as very slim.

The exception to this latter group is Malaysia in as far as there is evidence that it is likely to host a number of integrated plants combining wafer fabrication capabilities with automated assembly and testing facilities. As I implied in the previous chapter, however, it may be sensible to view these developments in Malaysia in relation to the complimentary development of regional management headquarters and design centres, in Singapore (and possibly Hong Kong). By so doing we might conclude that the regionally based sets of dependent relations and transactions within which the 'Malaysian' semiconductor industry has been entangled for some years, is likely to persist for some time to come.

Beyond the 'new' international division of labour

For those who have followed our argument to this point, it should now be clear that the historical details of the globalisation of the US semiconductor industry poses serious explanatory problems for the NIDL thesis. It is not my contention, however, that the thesis was totally inadequate from the beginning and therefore led students of economic development down analytic blind alleys. While the thesis was undoubtedly flawed, it still helped us to grasp many of the salient features of an earlier phase in the globalisation of certain types of manufacturing industry. Its problem now is that it is in danger of being rendered obsolete by history, though its 'obsolescence' is, in part, a product of the theoretical 'silences' endemic to the thesis, which have become increasingly exposed as the world economy has continued to evolve. But in what ways was the NIDL thesis relatively adequate to the task of helping us gain some explanatory purchase on the earlier phases of the globalisation of semiconductor production? And what are these 'silences' that need to be recovered from our previous discussion (in Chapter two) and in the light of our empirical investigations, elaborated or supplemented here?

The NIDL thesis was important in helping us understand many of the imperatives which led to the movement offshore of semiconductor assembly processes in the period from the early 1960s to about the mid-1970s. Although valorisation problems were perhaps not as significant initially for semiconductors as they were for other industries, the need to reduce costs and increase productivity in the face of rising competition (from Japan) were undoubtedly central impulses. Additionally, the ability to reduce costs by taking advantage of a major 'comparative advantage' which Third World societies possessed, large pools of unemployed or underemployed unskilled labour, was also a central impulse. The fact that these labour forces often existed in the context of national states which were prepared to legislate massively in favour of foreign capital and against the interests of workers was a significant contingency that also proved attractive to semiconductor manufacturers. Finally, the fact that semiconductor companies until the mid-1970s invested overwhelmingly in Third World countries only in the least technologically advanced labour processes, such as assembly, and not in full production capabilities, and that state-of-the-art technology generally was not used, is again consistent with the tenets of the NIDL thesis. For reasons such as these, then, the scepticism that foreign investment in export-oriented 'intermediate' manufacturing processes would do little to assist the genuine development of the host societies, seemed at first, to be reasonably well founded.

While elements of the NIDL thesis may continue to have a certain salience where foreign manufacturers currently are beginning to invest in Third World countries for the first time,[1] our discussion of the globalisation of semiconductor production has served to highlight a number of hiatuses which appear to have had serious analytic consequences:

1. In seeking to explain the spatial shifts in manufacturing investment that have occurred over the past quarter century or so, the NIDL thesis placed considerable emphasis on the problems of

valorisation that seemed to have become endemic to core economies. While this concern with the problems of the production of surplus value was an important corrective to much neo-Marxian development theory which, as I argued in Chapter two, tended to stress the realisation question, and hence the significance of commodity circulation, the NIDL theorists seem to have erred too far in the opposite direction. At a number of junctures in this study, we have seen how changes in the structure and location of markets have been centrally important for both the initial impulse towards the globalisation of semiconductor production and for the nature of the labour processes and production specialisations that have been implanted in particular locations. We suggested, for instance, that the changing structure of the US semicondutor market during the 1960s (from military to commercial customers) was in important factor facilitating the initial shift of assembly processes offshore. Additionally, we argued that military and subsequently EEC commercial markets was an important element propelling investment in Europe, including investment in technologically advanced production processes such as wafer fabrication. Finally, in the previous chapter we suggested that while the existence of local or regional markets was not a significant reason for the initial attraction of semiconductor investment to East Asia, the maintenance and indeed technological upgrading of plants in certain East Asian locations in recent years, has partly been associated with the expansion of market opportunities there.

2. In Chapter two, I suggested that a serious problem with the NIDL thesis, as with much of the tradition of analysis out of which it emerged, was its capital-logic approach to the explanation of peripheral industrialisation. Our account of the spatial dispersal of semiconductor production has highlighted how limited, and indeed analytically incorrect such an approach can be. Throughout the study I have emphasized how factors internal to the 'host' societies themselves can operate to shape the nature of the production system that results from foreign investment, and thus the extent to which it does or does not assist in the genuine development of the territorial unit in question. Of particular significance in this regard, has been state policy. Highlighting a few examples, we can point to the anti-trust policies of the US government forcing AT & T to license the technology for its new invention (the transistor) to other companies, or the same government's liberal tariff policies which encouraged the original offshore investment in low value-added assembly processes. Similarly, as I argued in Chapter six, EEC tariff policies were a

major inducement to US companies to invest in Britain in high value-added processes such as wafer fabrication.

While it has been evident, as the NIDL theorists argue, that national states have encouraged foreign investment through advantageous fiscal policies, the provision of transport and telecommunications infrastructure, the reproduction of low-wage labour markets by means of repression, or in the case of Hong Kong, subsidies, it is now clear that particular states have also intervened in other ways. Such intervention has not been at the 'behest' of transnational capital, but rather has been a product of their own development priorities. Some of the East Asian states, for instance, have effectively blocked foreign imports by means of prohibitive tariffs while encouraging investment in domestic manufacturing industries in the context of a development strategy which consciously privileges particular sectors. South Korea has been a prime example of such state-led industrialisation, so much so that it has caused one Marxist development economist to suggest that its success has been 'just as much a triumph of state capitalism as [were] the achivements of the first Five Year Plans in the Soviet Union or the People's Republic of China'. (Harris 1987: 145).

While South Korea, together with Taiwan, have been good examples of national states intervening to encourage the growth of an indigenously owned semiconductor production capacity, our study has highlighted additional ways in which state intervention has created conditions conducive not to investment in unskilled assembly processes, but rather to investment in relatively high-skill, capital-intensive production processes. Of particular note here has been the Singaporean government's deliberate attempt to force up labour costs in the knowledge that its labour market could supply the engineers and technicians necessary to ensure that foreign semiconductor and electronics corporations would not only continue to manufacture there, but on a restructured basis incorporating technologically advanced labour processes. Similarly, while the Hong Kong government has failed to support R & D provision for local electronics companies (thus, as I suggested in the previous chapter, risking the future of the entire industrial sector), it has created a system of higher education capable of reproducing the skilled labour power on which much of the territory's semiconductor industry now relies.

3. In Chapter two, I criticised the NIDL thesis for its excessive, indeed almost obsessive, concentration on the significance for globalisation of supplies of cheap unskilled labour at the

periphery of the world system. I suggested that, in theoretical terms, this pointed to a view of peripheral industrialisation which emphasized the opportunities for the creation of absolute surplus value at the expense of recognising other possibilities. I went on to argue that even if supplies of cheap unskilled labour were a major inducement for companies seeking to locate part of their production process offshore, this need not rule out the possibility that over time, and for various reasons, they might gradually shift to what we might call, loosely following Lipietz (1986), 'regimes' of relative surplus value creation. Under such circumstances, I suggested, one would expect rising wages to be a real possibility, including for the manual labour force.

Our account of the growth of semiconductor production in the developing societies of East Asia is an example of this contention. While, admittedly, some of the territorial units in the region have continued to be locked into regimes of absolute surplus value creation, at least as far as 'their' semiconductor industries are concerned, others have not. The gang of four are cases in point, though as I have suggested previously, it seems as though we shall soon have to cite Malaysia as an additional example.

In Chapters four, five and seven, I have outlined some of the reasons why these shifts to regimes of relative surplus value creation have occurred as part of the evolution of US semiconductor production in the region. I have suggested that while the changing structure and imperatives of the global operation of the industry in the context of world economic crisis have been important, the role of the national states themselves, as well as other processes internal to those societies, also have been crucial. I have addressed the significance of state policy in the previous point, but particularly important among the latter processes has been the emergence in South Korea, Taiwan, and Hong Kong of locally owned electronics and semiconductor production complexes. These have been important phenomena, in this sense, because they have provided contexts in which high-grade components, specialised services, and experienced engineering and technical labour power have been successfully reproduced. Consequently US (and other) semiconductor corporations have been able to invest in more advanced technologies and thus higher value-added products, in the reasonably sure knowledge that the social, technical, scientific, and physical support necessary for high yields under such circumstances, is already present in these particular territorial units.

4. While NIDL theorists have been largely negative in their assessments of whether foreign manufacturing investment stimulates

local industrial initiatives, they seem not to have recognised that the possibilities here may differ from one sector or branch to another, say as between textiles and electronics, and from one territorial unit to another. Our discussion of semiconductor production in the Philippines, for instance, pointed to the link between foreign investment and the growth of locally owned sub-contract assembly and other plants. Similarly, and much more spectacularly, however, in South Korea, Taiwan, Hong Kong, and Scotland, we have seen the rise of locally owned production complexes, manufacturing across a range of products and supplying many specialised services. Of additional significance is the fact that in the first three examples, locally owned electronics production (as well as that of other commodities, garments, and toys for instance, in Hong Kong) is now a much more important contribution to economic growth than is foreign-owned production.[2]

5. A criticism which is threaded in between a number of the comments made above, is the ahistorical manner in which the NIDL thesis has been elaborated. By arguing for the need to examine the changing dynamics of globalisation and the articulation of world-system factors with those internal to given societies over time, I have implied that it is simply not good enough to both begin and terminate one's analysis at the moment of initial manufacturing investment; at the moment the factories are in place and a production system begins to emerge. Unfortunately, however, this is precisely what the NIDL theorists have tended to do. In effect they have treated the territorial units as if they were *tabulae rasae* prior to the moment of foreign manufacturing investment. As Cohen (1987b) has argued, this is a major failure which negates, for instance, the entire experience of imperialism and the structural changes wrought by it. Thus recent scholarship has shown how important social structural changes in Korea and Taiwan under Japanese imperialism were to their subsequent industrialisation (e.g. Cummings 1987; Koo 1987; Gold 1986). Similarly, in Chapter five, I indicated how important both world system factors and pre-existing forms of economic activity were to Hong Kong's industrialisation in the 1950s.

An additional consequence of this absence of an historical dimension to the NIDL theorists' work is that they have failed to identify and/or anticipate the emergence of sub-global international divisions of labour within certain sectors or branches. As I have shown in Chapter four, the rise of a specifically regional division of labour within US semiconductor production in East Asia is a case in point. Our ability to identify such sub-global divisions of labour is rather important (this study suggests), precisely because they may be one of the factors leading to an increasing 'relative autonomy' of the development trajectories of certain territorial units, whilst further constraining the possibilities for others.

6. The final comment I wish to make here is a corollary of my previous point. Not only has the NIDL thesis been developed ahistorically, but it remains far too abstract to assist us to comprehend the actual economic, social, and political processes as they emerge and articulate within particular factories, production systems, and territorial units. While the NIDL thesis has undoubtedly been of help to us in identifying and assessing the global dynamics behind the spatial dispersal of semiconductor production, it has not been much use in helping us understand why there should have been investment in a particular territorial unit rather than another; why a particular labour process should be implanted say in Hong Kong rather than the Philippines, why semiconductor production has been able to be harnessed as part of a national development strategy more effectively in Singapore, South Korea, or Scotland rather than in Indonesia or Thailand. But it is precisely these sorts of issues that we need to comprehend if we are ever to get an adequate purchase on real development processes as they impact on actual societies at particular historical moments.

The world economy is in flux; and with it, what just ten years ago appeared as a 'new' international division of labour, has already been superseded. In this section I have highlighted some of the reasons for that supersession as far as semiconductor production is concerned. As with urban and regional theory in the 1970s, perhaps certain strands of development theory have moved too quickly towards theoretical closure. We clearly need general theorising to guide our investigations into the economic and social transformations of our age. The global economy appears to be changing so fast, however, that those guides must be transitional and therefore heuristic. They need to be constructed from concepts which 'sensitize' us to a range of empirical possibilities if they are to be at all useful for the purposes of explanation, and as aids to practice in what is an increasingly turbulent world. But issues such as these are for discussion in places other than the current volume. For now, I want to raise once more the question of high technology as a new mode of industrialisation.

Development and the new mode of industrialisation

I began this book by suggesting, tentatively, that electronics and other forms of high technology production might constitute the basis of a new mode of industrialisation. In this final section, we return to this issue and examine, in the light of our investigations, what the significance of this notion might be.

We have seen from our study of the semiconductor industry that while technological progress tends to undermine the importance of unskilled labour in total production costs, it does not necessarily mean that companies seeking to reconstruct their operations will pack up and move out of locations where cheap unskilled labour was presumed to be their principal asset. A corollary of this seems to be that it was wrong of the NIDL theorists and others to assume that where cheap unskilled labour appeared to be a major 'comparative advantage' of a national or local economy, then manufacturing investment would always be designed to exploit that resource. The assumption seemed to be that the existence of cheap labour in Third World societies would operate as a disincentive for companies to invest in high technology forms of production. Our study has shown that, in certain instances, these assumptions have not been borne out in practice. We have advanced a number of suggestions as to why these developments, in places such as the gang of four, have been possible, ranging, for instance, from the changing structure of the world economy and markets and corporate priorities in those contexts, to the development strategies of the national states in question. The additional issue that confronts us, however, is whether there is something peculiar about electronics and other high-technology forms of production which facilitates the developments highlighted above.

In the opening chapter, I outlined four elements which appeared to distinguish electronics industries from those on which earlier rounds of industrialisation had been based. From among those elements it would appear to be the distinctive function of knowledge that was particularly important for our current purposes. I suggested in the first chapter that for companies in an industrial sector such as electronics, so dependent as they are on technological innovation for their competitive edge, knowledge, embodied in scientific and technical labour power, becomes perhaps the principal factor of production. In Chapters four to six I showed that where territorial units can produce and continue to deliver a labour force with sufficiently advanced technological skills, then it becomes possible for them to shift from regimes of absolute to regimes of relative surplus value creation, and hence put in place one of the preconditions for genuine development. However, as I have argued throughout this text, the ability to reproduce scientific and technical labour power, though a necessary factor, is still an insufficient condition for a manufacturing economy to move technologically upstream. Without the other factors, both internal and contingent, such developments are unlikely to occur. Thus, for instance, although the Philippines has a relatively advanced tertiary education system, it has been unable so far to capitalise on this resource, and its electronics industry remains largely locked into a regime of absolute surplus value creation.

The electronics, or more generally high technology mode of industrialisation, then, is clearly not the solution to the development problems of the majority of Third World societies. While it has been crucial in certain societies, such as the gang of four, its record in others where it has taken root, is by no means as clearly positive. In many of those other locations – Indonesia, the Philippines, Thailand perhaps – it has been, if anything, the other elements of the mode of industrialisation that have had the more far-reaching consequences. The electronics industries in those and similar societies have remained locked into their position at the cheap labour-intensive, low value-added end of the international division of labour. It is in those societies, perhaps more than in others, that 'their' electronics industries with their polarising social and technical divisions of labour, have contributed more to the massive relative inequalities typical of so many less-developed capitalist societies than they have to development and rising living standards.

Even if the track record of this new mode of industrialisation, when judged by its capacity to assist development, differs significantly as between those societies affected by it, what of the bulk of the Third World that has yet to be touched by it? As Castells has recently pointed out, the 'real cleavage' may not be so much between societies differentially affected by the new mode of industrialisation, but rather

> between countries integrated in the international structure of production and those excluded from it because they do not offer the minimum conditions to incorporate up-to-date technologies necessary to compete in an interdependent world regardless of the cost of labor used in the production process.
>
> (Castells 1988: 16)

Thus for those territorial units which for whatever reason are unable to offer foreign electronics corporations anything but cheap, unskilled labour, they are now, in the context of the current world system, unlikely to be even touched by the new mode of industrialisation, never mind become recipients of some of its benefits. It is for this, among other reasons, that we must resist the temptation to assume that the gang of four, for instance, with their rising living standards, based as they are, in part, on semiconductor and other forms of electronics production, constitute a viable model for development elsewhere across the globe. The 'Third World' remains very much with us and therefore, as Lipietz (1987: 93–6) and Worsley (1984) have reminded us, so does the 'old', perhaps very old, international division of labour.

Notes

1 A new mode of industrialisation

1. Throughout this study I adopt the usage, by French political economists, of the terms industrial sector and branch. The term 'sector' refers to an entire industry, such as electronics or automobiles. The term 'branch', however, refers to a sub-set of products and production activities within an industrial sector. Examples within the electronics sector, for instance, would be computers or semiconductors; and in automobiles, cars or trucks (cf. Palloix 1977; Aglietta 1979).
2. The International Sociological Association's Research Committee on Urban and Regional Development, at least since 1982, has emerged as one important forum for this type of work. More recently, but particularly with regard to the Pacific Rim, institutes have emerged at the Universities of California (Los Angeles and San Diego campuses) and Sydney to sponsor transnational, multidisciplinary research of the sort indicated.
3. For the purposes of this study, the term 'East Asia' will be taken to include all the countries normally grouped under that rubric. However it will specifically exclude Asia's economic and socialist giants, Japan and China respectively.
4. By 'growth of transnational corporations' I refer not merely to the expansion of the activities of existing transnationals, but also to the increasing 'transnationalisation' of formerly national companies. It is consistent with contemporary restructuring that national companies in core economies frequently must adopt the 'global option' in order to survive in increasingly competitive domestic and world economies. This was precisely the situation that confronted US semiconductor companies, for instance, at the point they first began to internationalise their production in the early 1960s (see Chapter 3).

2 The international division of labour, industrial change, and territorial development: Theoretical and methodological issues

1. A recent exception here is the work of Richard Batley. Working out

from both dependency theory (particularly the version associated with Cardoso) and urban theory (particularly some of Castells' earlier work) and incorporating insights from organisation theory and the 'urban managerial' tradition in British sociology, he has developed a suggestive account of urban development (particularly with regard to housing policy) in Brazil. See Batley (1982, 1983).

2. In at least one contribution to the world-systems literature, however, Wallerstein (1978: 222) has insisted that the semiperiphery concept refers only to state processes. See, however, the extended discussion by Arrighi and Drangel (1986).
3. While it is the case that the rate of growth in foreign direct investment has been higher in core countries than peripheral or semiperipheral ones, in the manufacturing sector, the rate of increase has been higher in the Third World (see Slater 1985, Chapter 2).
4. For accounts of the internationalisation of financial and commercial capital, see Daly and Logan (1986) and Thrift (1987) respectively.
5. In many of the 'newly industrialising countries' which have substantial domestic markets (e.g. Brazil, Argentina), foreign manufacturing companies have not engaged in export-oriented production to the extent that Fröbel *et al.* suggest. See Schiffer (1981) for details.
6. In spite of the existence of informal sectors, however, there has been a rise in real wages in the formal sector of 'developing' countries to the tune of 4.4 per cent, per annum, 1950–75 (Schiffer 1981). The point here is that without the existence of informal sectors, and the limited or 'semi' proletarianisation which they imply (cf. Wallerstein 1983), the rate of increase in real wages in the formal sector probably would have been greater.
7. The significance of the 'law' of the tendency of the rate of profit to fall for an analysis of the origins of the internationalisation of industrial capital, has recently been critically questioned by James O'Connor (1981, 1984).
8. By 'genuine' development, I mean development that significantly improves the economic and social well-being of the overwhelming majority of the population in the given territorial unit, and that this development is relatively permanent (in as far as history can create anything that is 'permanent'). This is a minimum definition, but could be expanded to include, for instance, notions of a democratic polity.
9. Marx's concept of 'absolute surplus value' refers to the situation where a capitalist seeks to increase surplus value (and hence capital accumulation) by means of increasing the ratio of surplus labour time (that part of the work period when surplus value is created) to necessary labour time (that part of the work period when the worker produces the value necessary for his or her own, and the family's subsistence needs). Under conditions of absolute surplus value creation, the ratio of surplus to necessary labour time is increased by means of forcing down wages,

forcing people to work harder for the same wage and/or lengthening the working day. The creation of absolute surplus value is associated, therefore, with the 'super-exploitation' of the labour force, or to what Lipietz (1987) refers to as 'bloody taylorisation'.

The concept of 'relative surplus value', on the other hand, refers to a situation where surplus value is expanded without necessarily having recourse to 'super-exploitation'. This is done by increasing productivity by means of the application of (new) technology. Under a regime of relative surplus value creation, it is quite possible for wages to increase at the same time as increases in surplus value occur. The classic discussions of absolute and relative surplus value are contained, of course, in the first volume of *Capital* (Marx 1967: Parts III and IV). However, also see Harvey's (1982: Chapter 1) useful account.

3 Semiconductor production: Labour processes, markets and the determinants of globalisation

Parts of this chapter were co-authored with Allen Scott and originally published in *The Development of High Technology Industries*, edited by Michael Breheny and Ronald McQuaid, London: Croom Helm, 1987.

1. For a theoretical and empirical demonstration of why the necessary internal relations of any industrial branch are embodied within labour processes and their technical evolution, see Marx's classic account in the first volume of *Capital* (Marx 1967, Chapters 13–15).
2. Manuel Castells (1987), disputes the argument that strong unionisation in the Northeastern states was one of the reasons why semiconductor production did not emerge there, initially, to the same extent that it did in California. He suggests, rather, that semiconductor companies have developed production (and we might add, labour relations) systems that tend to circumvent the possibilities for unionisation, wherever they locate. He suggests this is true even when they locate in areas of militant union activity, such as Scotland. This strikes me as an interesting hypothesis, but one that is still in need of empirical verification.
3. Clearly I am arguing here for a dialectical understanding of the significance of university research and educational programmes for the emergence of high technology industry. See Noble (1977) for a similar understanding of the relationship between universities and the 'first' industrialisation of the United States.
4. In developing this notion I have been strongly influenced by the work of Allen Scott. For his own work on Silicon Valley, see Scott and Angel (1986).
5. Information from Lenny Siegel, Director, Pacific Studies Center, Mountain View, California.

4 East Asia: The emergent regional division of labour

1. 'Merchant' producers are those semiconductor companies who sell all of their product on the open market. 'Captive' companies, on the other hand, are in effect divisions of electronics systems corporations, and the bulk of their product (usually in excess of 75 per cent) is used by the parent company.

2. This is the 'second-sourcing' phenomenon designed to minimize the disruption of the global production chain associated with strikes, political turmoil etc. in particular locations at particular historical moments.

3. The retention of this arrangement through to the present would seem to suggest that even the most 'advanced' of East Asia's electronics production complexes have still not developed to the extent that they can guarantee American manufacturers the types of controls necessary for the high-yield production of such a technologically sophisticated product as semiconductors. Licensing arrangments, however, have developed between US and Korean and Taiwanese companies (see later sections of this chapter for details), though I am aware of only one joint-venture operation, and that is between South Korea's Goldstar Semiconductor and the US holder of 44 per cent of its equity, AT & T. (*Global Electronics Information Newsletter*: 16, November, 1981).

4. Note, however, that Motorola has never had assembly processes in its Hong Kong plant.

5. In the case of Motorola, their Hong Kong design centre is one of their regional centres (the others being in Switzerland and Japan). Zilog has set up a design centre without having, as yet, a manufacturing operation; and Siliconix's design centre is a joint venture with a local Hong Kong firm.

6. In 1986 the Italian–French company, SGS, set up a wafer fabrication plant in Singapore.

7. Note that the Indonesian semiconductor industry has never expanded beyond the two plants established in the mid-1970s. Indeed a semiconductor presence in Indonesia appears on the verge of elimination. Fairchild closed its plant there in 1986 and National Semiconductor has recently cut its workforce by 60 per cent (*Global Electronics*: 62, February 1986).

8. Hyundai reputedly had to close both its design centre and wafer fabrication plant in Silicon Valley as US engineers would not tolerate its autocratic management style (*Fortune*, 16 March 1987). Other Korean companies operating in California have attempted to circumvent this problem by recruiting, as far as possible, Korean-American engineers (personal communication from Lenny Siegel, Director, Pacific Studies Center, Mountain View, California).

9. Samsung, however, has begun to produce the more technologically

advanced 256K RAMs in commercial quantities, and hopes to develop the 1 megabyte RAM by the end of the 1980s (*Electronics*, 16 September 1985).

10. That is, integrated circuits with up to 148 wire leads each.
11. At least one scholar has argued that the relative absence of an indigenously owned electronics sector in Singapore, poses no problems for the future development of the industry there (Lim 1987).
12. Information from fieldwork interviews at National Semiconductor, Penang, Malaysia, April 1988.

5 Hong Kong: The making of a regional core

1. No reliable data on the Hong Kong economy prior to the Second World War is available, but all the standard accounts of Hong Kong's development (e.g. Szczepanik 1958; Chou 1966; Riedel 1972; Lin and Ho 1980; Cheng 1985) are in accord with this statement.
2. By 1975 Hong Kong had overtaken Italy as the world's largest exporter of clothing (Lin and Ho 1980: 34).
3. As a consequence of crisis and overproduction in the international semiconductor industry in the 1980s (cf. Ernst 1987), Fairchild substantially reduced its Hong Kong operation in 1983 and finally closed its factory in 1986, shipping production equipment to its South Korean facility.
4. Some indicative figures for the proportion of employment provided by manufacturing industry (1986) are:

	%
Hong Kong	34
Taiwan	32.5
Japan	25
Singapore	25
South Korea	22.5
Malaysia	14.6
Indonesia	10.4
Philippines	9.5
Thailand	8.3

Apparently, of all East Asian countries, only North Korea has a higher proportion of its workforce (53.2 per cent) employed in manufacturing industry (BBDO 1987: 3).

5. Readers who have never been to Hong Kong and have an image of factories as buildings which occupy space horizontally, may wonder where the land for 1,200 electronics (not to mention thousands of others) factories in a territory of 1,000 square kilometers and 5.6 million people, comes from. The answer, of course, is that like the population itself, factories in Hong Kong occupy space vertically!
6. Unfortunately, because of the absence of a law of disclosure in Hong

Kong, and the unwillingness of local industrialists to respond to questions about investment, it is almost impossible to generate reliable data on investment in the locally owned electronics sector. Consequently data that could confirm this point cannot be presented at this time.

7. This decline has probably been associated with the crisis in the worldwide consumer electronics industry, associated in particular, with the saturation of the VCR (domination by Japanese producers) and micro-computer markets by the early to mid-1980s. On the general significance of this crisis for both Japanese and US producers, see Ernst (1987).
8. Based on unpublished data drawn from the Reports on the Surveys of Overseas Investment in Hong Kong's Manufacturing Industry, 1981 and 1987. Available from the Hong Kong Department of Trade, Industry and Customs.
9. Given that Table 5.8 does not include data on the minor EEC markets (for Hong Kong-produced electronics goods), this figure must *underestimate* the market dominance of the US and EEC.
10. In addition to the US companies listed in Table 5.11, NCR and Silicon Systems both had semiconductor assembly plants in Hong Kong until the early 1980s (Grunwald and Flamm 1985; UNCTC 1986).
11. A 'total' account of Hong Kong's development is in preparation (Henderson 1989).
12. The first EPZ/FTZs in the region were not set up until 1966 (in Taiwan and South Korea). Those in Malaysia and the Philippines did not begin operations until 1972 and 1973 respectively (Grunwald and Flamm 1985: 78). The penetration of US semiconductor investment into the latter three countries was directly associated with the emergence of their EPZs. The only other territorial unit in the region which offered the same free port advantages as Hong Kong, was Singapore. Like Hong Kong its free port status stemmed historically from the same role in the world economy, as a British colonial entrepôt.
13. Significant industrialisation in Taiwan and South Korea, for instance, did not begin much before the early 1960s (see Gold 1986; and a number of the essays in Deyo 1987).
14. General accounts of the processes and significance of habituation to factory labour have been provided by Henderson and Cohen (1979, 1982b) amongst others.
15. Accounts of the cultural problems (resistance) associated with attempts to convert peasants or agricultural labourers into factory workers are legion, and refer to a variety of racial and ethnic groups, national contexts, and to both sexes. For instance see Cohen (1980) on Africa and Heyzer (1986) on Southeast Asia. The classic historical accounts, however, remain those by Thompson (1967) and Braverman (1974).

16. This has not always been the case. There were apparently a number of strikes during the nineteenth century, but particularly in the 1920s when a series of seamen and dockworker strikes culminated in the fifteen-month general strike of 1925–6. (Turner *et al.* 1980: 19–20; England and Rear 1981: 124–33).
17. Probing the formation of workforce consent could only be achieved by the use of ethnographic techniques, which would need to include extensive periods of participant observation within the workforce. While three ethnographies of workplaces in Hong Kong have been completed (Djao 1976, 1978; Salaff 1981; Chiang 1984), none of them used the formation of workplace consent as part of their theoretical problematic.
18. I am attempting to address some of these questions elsewhere (Henderson 1989).
19. Sun's book is currently available only in Chinese. I am grateful to Ho Shuet Ying for providing me with a detailed exposition of Sun's arguments. For a brief English language summary, see Jaivin (1987).

    I am acutely conscious of the fact that in other historical and structural contexts, Chinese workers have been as militant as any others. I have already referred to the labour militancy in Hong Kong during the inter-war period, for instance, and thanks to the work of Jean Chesneaux (1968) among others, we now know a great deal about working-class militancy in China itself. In no sense, then, should my comment be seen as a culturalist explanation for the present lack of militancy among the Hong Kong workforce. Rather my position is that culture (as a changing, not invariant form) *may* be *one* element of the explanation. We cannot even begin to understand its significance, however, unless (a) we approach the question by means of a theoretical framework which recognises that culture is articulated with other structural elements and that *the articulation changes over time* and (b) that empirical work is required to identify its significance in particular institutional contexts at particular historical moments.
20. In my interviews with the managers of US semiconductor plants in Hong Kong, this comment was invariably given in response to the question of why assembly workers in the factory were women.

6 Scotland: The European connection

1. It should be borne in mind, however, that European semiconductor production as a whole, is small by world standards. A recent estimate suggests that Europe has captured only 7 per cent of the world market, compared with 53 and 39 per cent respectively for the USA and Japan (*The Economist*, 24–30 November 1984). The Scottish Development Agency is more optimistic, putting the figure at 21 per cent in 1983 (Locate in Scotland, 1983: 89).
2. This and the following section draw extensively on information gained

from interviews with executives of American semiconductor companies, trade union officials, representatives of the Scottish Development Agency etc. The interviews were conducted in June and July, 1985.

3. Indeed, since the research on which this chapter is based was completed, one of the companies, General Instrument, has closed its Scottish plant with a loss of 210 jobs.
4. Data compiled from Scottish Development Agency (1984a, 1984b) and updated from SDA records of April 1985.
5. Data drawn from an SDA survey of June 1983 and compiled by Firn and Roberts (1984: 300). Foreign (i.e. predominantly American) control over the electronics industry in Scotland is also significantly higher than in Britain as a whole. In 1977, for instance, foreign companies accounted for 45 per cent of electronics employment in Scotland, but only 26 per cent in the UK. (Horn and Hetherington 1982, Table 2: 16).
6. National Semiconductor and Burr-Brown are now the only US semiconductor firms to maintain design centres in Scotland. (Compare the data in Figures 3.2 and 4.2).
7. Hughes Aircraft, for instance, who are beginning an expansion programme, anticipate that by 1989 they will have a total workforce of 1,200, of whom 60 per cent will be engineers or technicians.
8. Feminist analyses of women's work are replete with empirical detail which substantiates this point. See for instance, Pollert (1981) for factory work, Barker and Downing (1980) and Downing (1981) for office work, and a number of the essays collected in West (1982) for a variety of work situations.
9. At best this strategy to circumvent 'bad' work habits is full of contradictions. Compare, for instance, the problematic (for management) experience at the Ford Halewood plant (England) when teenage workers were introduced to assembly-line labour (Beynon 1973: 139–40).
10. Information from John Langan, ASTMS organiser, Glasgow.
11. Wages for manual employees at Hughes are in fact only about average for their local labour market (Glenrothes/Kirkcaldy). They score heavily, however, with a 4½-day, 39-hour working week and four weeks paid leave a year (5 weeks after 5 years' service, 6 weeks after 10 years') for all employees, whatever their grade.
12. The US semiconductor houses AMD and Zilog reputedly opted for Ireland over Scotland as their European production bases in early 1985, because of the less-advantageous grants they were eligible for in Scotland, subsequent to the 1984 reorganisation of regional aid (see Figure 4.2).

7 Prospects for globalisation and development

1. Information from *Global Electronics Information Newsletter* (various issues), *Global Electronics* (various issues), *Electronics* (various issues), Hong Kong Productivity Centre, *Electronics Bulletin* (various issues), *San Francisco Examiner* (9 November 1986) and Ernst (1987).
   Strategic alliances are part of the restructuring process of a number of industries, in addition to semiconductors. For details see Hill (1987) on the automobile industry and Ohmae (1985) for a general account.
2. Jacob Retinoff, Chairman and President of Indy Electronics, an American assembly and test subcontractor with a plant in Scotland; quoted by Hargrave (1985: 41).
3. Singapore's viability is also based on supplies of water and gas from Malaysia. Any future conflicts between these states could place the Singaporean economy in serious jeopardy.
4. For a general, if 'early', account of the growing economic interdependencies of Hong Kong and China see Youngson (1983).
5. Not surprisingly in these circumstances, although Government employment generally is at a standstill, the police force continues to expand at the rate of about 300 additional officers a year (with 30,000 officers it is already almost as large as that of New York City). The police have apparently developed a policing strategy based on the assumption that the territory's economy will collapse around 1992, and that serious conflict will ensue.
6. R & D investment by Hong Kong firms is unlikely, partly because of their small size and hence lack of capital, but also because of the political and therefore economic uncertainties associated with the Colony's absorbtion by China in 1997.
7. South Korea is the only one of the gang of four with a potentially large domestic market. It should be remembered that with the possible exception of Sweden, no country has yet achieved advanced industrial status without possessing a substantial domestic market.

8 Semiconductors, development, and the changing international division of labour

1. With regard to technology transfer, China, for instance, may be a case in point. Although systematic research on the nature and consequences of foreign manufacturing investment in China is not yet available (but see the forthcoming report by Bianchi *et al.* 1988) informal observations suggest that the experience there may be similar to that of other East Asian countries in the 1960s and 1970s. Three examples come to mind. In September 1986 on a lecture-study tour to Beijing, I had the opportunity to visit two joint-venture plants, one a computer plant in which Hewlett-Packard was the foreign partner and the other an automobile factory (assembling jeeps) where the arrangement was then

with American Motors (now part of the Chrysler empire). In the first case computers for the domestic market were assembled from 'kits' imported from the United States. They embodied no local content, nor were there plans to incorporate Chinese-produced components in the future. Yet the manager with whom I spoke repeatedly referred to 'technology transfer' as one of the benefits China was reaping from Hewlett-Packard's involvement. When I asked what technology was in fact being transferred, the manager paused for thought, and eventually suggested 'management skills'. In the case of the automobile joint-venture, local content was again very low (apparently below 5 per cent), though there the intention was to increase the use of domestic components over the next few years. What was particularly striking for anyone aware of modern automobile production processes, however, was the nature of the assembly line. In terms of the technology embodied, it seemed like something that Detroit would have discarded about 1960. In the same vein, one senior semiconductor engineer with whom I spoke in 1984 referred to the technology used in Hitachi's joint-venture plant in China as 'geriatric'.

2. Foreign companies are responsible for only about 12–15 per cent of manufactured commodities by value in Hong Kong, and about 15–18 per cent in Taiwan and South Korea.

Bibliography

Aglietta, M. (1979) *A Theory of Capitalist Regulation: The US Experience*, London: New Left Books.

Amin, S. (1974) *Accumulation on a World Scale*, New York: Monthly Review Press.

(1976) *Unequal Development*, New York: Monthly Review Press.

Armstrong, W. and McGee, T. G. (1985) *Theatres of Accumulation, Studies in Asian and Latin American Urbanisation*, New York: Methuen.

Aronowitz, S. (1974) *False Promises, The Shaping of American Working Class Consciousness*, New York: McGraw-Hill.

(1978) 'Marx, Braverman and the logic of capital', *Insurgent Sociologist* 8 (2 & 3): 126–46.

(1981) 'A metatheoretical critique of Immanuel Wallerstein's The Modern World System', *Theory and Society*, 10 (4): 503–20.

(1982) 'The end of political economy', pp. 139–200 in S. Aronowitz, *The Crisis in Historical Materialism*, New York: Praeger.

Arrighi, G. and Drangel, J. (1986) 'The stratification of the world economy: an exploration of the semiperipheral zone', *Review* X (1): 9–74.

Baggott, M. (1985a) 'Scotland's Silicon Glen', *120 Scottish Economic Development Review* 3: 16–19.

(1985b) 'The vital support industries', *120 Scottish Economic Development Review* 4: 14–17.

Baran, P. (1967) *The Political Economy of Growth*, New York: Monthly Review Press.

Baran, P. and Sweezy, P. (1966) *Monopoly Capital*, Harmondsworth: Penguin.

Barker, J. and Downing, H. (1980) 'Word processing and the transformation of the patriarchal relations of control in the office', *Capital and Class* 10: 64–99.

Batley, R. (1982) 'The politics of administrative allocation', pp. 78–111 in R. Forrest, J. Henderson and P. Williams (eds) *Urban Political Economy and Social Theory*, Aldershot: Gower.

(1983) *Power through Bureaucracy, Urban Political Analysis in Brazil*, Aldershot: Gower.

BBDO (1987) *A Regional Focus: Hong Kong and S. E. Asia*, Hong Kong: BBDO Ltd.

Bello, W., Kinley, D. and Elinson, E. (1982) *Development Debacle, The World Bank in the Philippines*, San Francisco: Institute of Food and Development Policy/Philippine Solidarity Network.
Berney, K. (1985) 'The four dragons rush to play catch-up game', *Electronics Week*, 6 May: 48–52.
Bernstein, A., DeGrasse, B., Grossman, R., Paine, C. and Siegel, L. (1977) *Silicon Valley: Paradise or Paradox?*, Mountain View: Pacific Studies Center.
Beynon, H. (1973) *Working for Ford*, Harmondsworth: Penguin.
Bianchi, P., Carnoy, M. and Castells, M. (1988) *Economic Modernisation and Technology Transfer in the People's Republic of China*, Stanford: School of Education, Stanford University.
Blackaby, F. (ed.) (1979) *Deindustrialisation*, London: Heinemann.
Bluestone, B. and Harrison, B. (1982) *The Deindustrialisation of America*, New York: Basic Books.
Bolton, K. R. and Luke, K. K. (1986) 'The socio-linguistic survey of language in Hong Kong', Mimeo, Department of English Studies and Comparative Literature, University of Hong Kong.
Borrus, M., Millstein, J. and Zysman, J. (1982) *International Competition in Advanced Industrial Sectors: Trade and Development in the Semiconductor Industry*, Washington DC: Joint Economic Committee, Congress of the United States, US Government Printing Office.
Braudel, F. (1980) *On History*, Chicago: University of Chicago Press.
Braun, E. and MacDonald, S. (1982) *Revolution in Miniature*, Cambridge: Cambridge University Press.
Braverman, H. (1974) *Labor and Monopoly Capital*, New York: Monthly Review Press.
Brecher, J. (1973) *Strike!*, San Francisco: Straight Arrow Books.
Breheny, M. and McQuaid, R. (1987) 'H.T.U.K.: the development of the United Kingdom's major centre of high technology industry', pp. 297–354 in M. Breheny and R. McQuaid (eds) *The Development of High Technology Industries: An International Survey*, London: Croom Helm.
Brenner, R. (1977) 'The origins of capitalist development, a critique of Neo-Smithian Marxism', *New Left Review* 104: 25–92.
Browett, J. (1985) 'The newly industrialising countries and radical theories of development', *World Development* 13: 789–803.
(1986) 'Industrialisation in the global periphery, the significance of the newly industrialising countries of East and Southeast Asia', *Environment and Planning D: Society and Space* 4(4): 401–18.
Burawoy, M. (1979) *Manufacturing Consent*, Chicago: University of Chicago Press.
Campbell, R. H. (1980) *The Rise and Fall of Scottish Industry*, Edinburgh: John Donald.
Cardoso, F. H. and Faletto, E. (1979) *Dependency and Development in Latin America*, Berkeley: University of California Press.
Castells, M. (1977) *The Urban Question*, London: Edward Arnold.
(1978) *City, Class and Power*, London: Macmillan.
(1983) *The City and the Grassroots*, London: Edward Arnold.

(1986a) 'The Shek Kip Mei syndrome, public housing and economic development in Hong Kong', *Working Paper* 15: Centre of Urban Studies and Urban Planning, University of Hong Kong.

(1986b) 'High technology, economic policies and world development', *BRIE Working Paper*, University of California, Berkeley.

(1987) 'The new industrial space: high technology manufacturing and spatial structure in the United States', Mimeo, Department of City and Regional Planning, University of California, Berkeley.

(1988) 'High technology and the new international division of labour', Paper to the International Institute for Labour Studies/Indian Council of Social Science Research Seminar on the Diffusion of High Technology and the Labour Market: Asian Experiences, New Delhi, March.

Castells, M. and Henderson, J. (1987) 'Techno-economic restructuring, socio-political processes and spatial transformation, a global perspective', pp. 1–17 in J. Henderson and M. Castells (eds), *Global Restructuring and Territorial Development*, London: Sage Publications.

Chase-Dunn, C. (ed.) (1982) *Socialist States in the World-System*, Beverly Hills: Sage Publications.

Chen, E. K. Y. (1971) 'The Electronics Industry of Hong Kong: An Analysis of its Growth', M.Soc.Sc. Thesis, University of Hong Kong.

Cheng, T. Y. (1985) *The Economy of Hong Kong*, Hong Kong: Far East Publications.

Chesneaux, J. (1968) *The Chinese Labor Movement, 1919–1927*, Stanford: Stanford University Press.

Chiang, S. N. C. (1984) 'Women and Work: Case Studies of Two Hong Kong Factories', M.Phil. Thesis, University of Hong Kong.

Chiu, P. K. Y. (1986) 'Labour Organisations and Political Change in Hong Kong', M.Soc.Sc. Dissertation, University of Hong Kong.

Chou, K. R. (1966) *The Hong Kong Economy: A Miracle of Growth*, Hong Kong: Academic Publications.

Cohen, R. (1980) 'Resistance and hidden forms of consciousness among African workers', *Review of African Political Economy* 19: 8–22.

(1987a) *The New Helots: Migrants in the International Divison of Labour*, Aldershot: Avebury.

(1987b) 'The "new" international division of labour: a conceptual, historical and empirical critique', *Migration* 1(1): 21–46.

Cooke, P., Morgan, K. and Jackson, D. (1984) 'New technology and regional development in austerity Britain: the case of the semiconductor industry', *Regional Studies* 18(4): 277–89.

Cooper, E. (1981) *Woodcarvers of Hong Kong*, Cambridge: Cambridge University Press.

Crawford, R. (1984) 'The electronics industry in Scotland', *Fraser of Allander Institute Quarterly Economic Commentary* 9(4): 78–94.

Cressey, P. (1984) 'Participation in the electronics sector: the Comco case study', *Working Paper*, Centre for Research into Industrial Democracy and Participation, University of Glasgow.

Cressey, P., Eldridge, J. E. T. and MacInnes, J. (1986) *Just Managing*, Milton Keynes: Open University Press.

Cummings, B. (1987) 'The origins and development of the Northeast Asian political economy: industrial sectors, product cycles and political consequences', pp. 43–83 in F. Deyo (ed.) *The Political Economy of the New Asian Industrialism*, Ithaca: Cornell University Press.

Daly, M. T. and Logan, M. I. (1986) 'The international financial system and national economic development patterns', pp. 37–62 in D. Drakakis-Smith (ed.) *Urbanisation in the Developing World*, London: Croom Helm.

Deyo, F. C. (1981) *Dependent Development and Industrial Order*, New York: Praeger.

Deyo, F. (ed.) (1987) *The Political Economy of the New Asian Industrialism*, Ithaca: Cornell University Press.

Dicken, P. (1986) *Global Shift: Industrial Change in a Turbulent World*, London: Harper & Row.

Djao, A. W. (1976) 'Social Control in a Colonial Society: A Case Study of Working Class Consciousness in Hong Kong', Ph.D. Dissertation, University of Toronto.

(1978) 'Dependent development and social control: labour-intensive industrialization in Hong Kong', *Social Praxis* 5 (3–4): 275–93.

Downing, H. (1981) 'Developments in Secretarial Labour: Resistance, Office Automation and the Transformation of Patriarchal Relations of Control', Ph.D. Thesis, University of Birmingham.

Edwards, R. (1979) *Contested Terrain*, London: Heinemann.

Ehrlich, P. (1988) 'Malaysia: fabrication hot spot?', *Electronic News*, 6 June: 18.

Emmanuel, A. (1972) *Unequal Exchange*, New York: Monthly Review Press.

England, J. and Rear, J. (1981) *Industrial Relations and Law in Hong Kong*, Hong Kong: Oxford University Press.

Ernst, D. (ed.) (1980) *The New International Division of Labour, Technology and Underdevelopment*, Frankfurt: Campus Verlag.

(1983) *The Global Race in Microelectronics*, Frankfurt: Campus Verlag.

(1985) 'Automation and the worldwide restructuring of the electronics industry: strategic implications for developing countries', *World Development* 13(2): 333–52.

(1987) 'US-Japanese competition and the worldwide restructuring of the electronics industry, a European view', pp. 38–59 in J. Henderson and M. Castells (eds) *Global Restructuring and Territorial Development*, London: Sage Publications.

Evans, P. (1979) *Dependent Development*, Princeton: Princeton University Press.

Evans, P. and Timberlake, M. (1980) 'Dependence, inequality and the growth of the tertiary, a comparative analysis of less developed countries', *American Sociological Review* 45(4): 531–52.

Firn, J. and Roberts, D. (1984) 'High-technology industries', pp. 288–352 in N. Hood and S. Young (eds) *Industry, Policy and the Scottish Economy*, Edinburgh: Edinburgh University Press.

Flynn, N. (1984) 'The dynamics of economic decline', Mimeo, Institute of

Local Government Studies, University of Birmingham.
Forrest, R., Henderson, J. and Williams, P. (1982) 'The nature of urban studies', pp. 1–14 in R. Forrest, J. Henderson and P. Williams (eds) *Urban Political Economy and Social Theory*, Aldershot: Gower.
Foster-Carter, A. (1978) 'The modes of production controversy', *New Left Review* 107: 47–77.
Frank, A. G. (1971) *Capitalism and Underdevelopment in Latin America*, Harmondsworth: Penguin.
(1980) *Crisis, In the World Economy*, London: Heinemann.
(1981) *Crisis, In the Third World*, London: Heinemann.
Fraser of Allander Institute (1985) *Fraser of Allander Institute Quarterly Economic Commentary* 10(4).
Friedman, A. (1977) *Industry and Labour*, London: Macmillan.
Friedman, M. and Friedman, R. (1981) *Free to Choose*, Harmondsworth: Penguin.
Friedmann, J. (1986) 'The world city hypothesis', *Development and Change* 17(1): 69–83.
Fröbel, F., Heinrichs, J. and Kreye, O. (1980) *The New International Division of Labour*, Cambridge: Cambridge University Press.
Gaffikin, F. and Nickson, A. (1985) *Jobs, Crisis and the Multinationals*, Birmingham: Birmingham Trade Union Resource Centre.
Gold, T. (1986) *State and Society in the Taiwan Miracle*, Armonk: M. E. Sharpe.
Goldstein, C. (1985) 'The necessity of invention', *Far Eastern Economic Review*, 21 November.
Goldstein, N. (1984) 'The women left behind: technological change and restructuring in the electronics industry in Scotland', Paper to the Workshop on Women and Multinationals, University of East Anglia.
Gregory, C. (1985) 'British Labor in Britain's Decline', Ph.D. Dissertation, Harvard University.
Grossman, R. (1979) 'Women's place in the integrated circuit', *South-East Asia Chronicle* 66: 2–17.
Grunwald, J. and Flamm, K. (1985) *The Global Factory*, Washington DC: The Brookings Institution.
Gutman, H. G. (1977) *Work, Culture and Society in Industrialising America*, Oxford: Basil Blackwell.
Habermas, J. (1973) 'What does a crisis mean today?: legitimation problems in late capitalism', *Social Research* 40(4): 643–67.
Hall, P. (1985) 'Technology, space and society in contemporary Britain', pp. 41–52 in M. Castells (ed.) *High Technology, Space and Society*, Beverly Hills: Sage Publications.
Hamilton, C. (1983) 'Capitalist industrialisation in East Asia's Four Little Tigers', *Journal of Contemporary Asia* 13(1): 35–73.
Hargave, A. (1985) *Silicon Glen: Reality or Illusion?*, Edinburgh: Mainstream Publishing.
Harris, N. (1987) *The End of the Third World*, Harmondsworth: Penguin.
Harvey, D. (1973) *Social Justice and the City*, London: Edward Arnold.
(1982) *The Limits to Capital*, Oxford: Basil Blackwell.

(1985) *The Urbanisation of Capital*, Oxford, Basil Blackwell.
Haug, P. (1986) 'US High Technology Multinationals and Silicon Glen', *Regional Studies* 20(2): 103–16.
Henderson, J. (1987) 'Semiconductors, Scotland and the international division of labour', *Discussion Paper 28*, Centre for Urban and Regional Research, University of Glasgow.
(1989) *The Political Economy of Hong Kong*, London: Routledge.
Henderson, J. and Castells, M. (eds) (1987) *Global Restructuring and Territorial Development*, London: Sage Publications.
Henderson, J. and Cohen, R. (1979) 'Capital and the work ethic', *Monthly Review* 31(6): 11–26.
(1982a) 'The international restructuring of capital and labour, Britain and Hong Kong', Paper to the International Sociological Association's Xth World Congress, Mexico City, August.
(1982b) 'On the reproduction of the relations of production', pp. 112–43 in R. Forrest, J. Henderson and P. Williams (eds) *Urban Political Economy and Social Theory*, Aldershot: Gower.
Heyzer, N. (1986) *Working Women in Southeast Asia*, Milton Keynes: Open University Press.
Hill, R. C. (1987) 'Global factory and company town, the changing division of labour in the international automobile industry', pp. 18–37 in J. Henderson and M. Castells (eds) *Global Restructuring and Territorial Development*, London: Sage Publications.
Ho, S. Y. (1986) 'Public housing', pp. 331–53 in J. Y. S. Cheng (ed.) *Hong Kong in Transition*, Hong Kong: Oxford University Press.
Ho. Y. P. (1986) 'Hong Kong's trade and industry, changing patterns and prospects', pp. 165–207 in J. Y. S. Cheng (ed.) *Hong Kong in Transition*, Hong Kong: Oxford University Press.
Hoogvelt, A. M. M. (1982) *The Third World in Global Development*, London: Macmillan.
Horn, M. E. and Hetherington, I. P. (1982) 'Overseas owned manufacturing establishments in Scotland: output, investment and employment', *Scottish Economic Bulletin* 24: 15–21.
Hymer, S. (1979) 'The multinational corporation and the international division of labour', pp. 140–64 in S. Hymer *The Multinational Corporation, A Radical Approach*, Cambridge: Cambridge University Press.
Ip, H. S. (1983) 'Hong Kong's Development: A Dependency Case?', M.Soc.Sc. Dissertation, University of Hong Kong.
Iscoff, R. (1986) 'Offshore assembly: a time of change', *Semiconductor International*, June: 96–102.
Jacobson, D., Wickham, A. and Wickham, J. (1979) 'Review of Die Neue Internationale Arbeitsteilung', *Capital and Class* 7: 125–30.
Jaivin, L (1987) 'The raw or cooked versions of being Chinese', *Far Eastern Economic Review*, 23 April: 43–5.
Jenkins, R. (1984) 'Divisions over the international divison of labour', *Capital and Class* 22: 28–57.
Johnstone, B. (1988) 'Microprocessors sharpen America's technological

edge', *Far Eastern Economic Review*, 7 July: 43–8.
Jomo, K. S. (1986) *A Question of Class: Capital, the State and Uneven Development in Malaya*, Singapore: Oxford University Press.
Katz, N. and Kemnitzer, D. S. (1982) 'Fast forward: the internationalisation of Silicon Valley', pp. 332–45 in J. Nash and M. P. Fernandez-Kelley (eds), *Women, Men and the International Division of Labor*, Albany: State University of New York Press.
Keller, J. F. (1981) 'The Production Worker in Electronics: Industrialization and Labor Development in California's Santa Clara Valley', Ph.D. dissertation, University of Michigan.
Koo, H. (1987) 'The interplay of state, social class and world system in East Asian development: the cases of South Korea and Taiwan', pp. 165–81 in F. Deyo (ed.) *The Political Economy of the New Asian Industrialism*, Ithaca: Cornell University Press.
Laclau, E. (1977) *Politics and Ideology in Marxist Theory*, London: New Left Books.
Lall, S. (1983) *The New Multinationals*, Chichester: John Wiley.
Läpple, D. (1985) 'Internationalisation of capital and the regional problem', pp. 43–75 in J. Walton (ed) *Capital and Labour in the Urbanized World*, London: Sage Publications.
Lau, S. K. (1982) *Society and Politics in Hong Kong*, Hong Kong: Chinese University Press.
Lebas, E. (1982) 'Urban and regional sociology in advanced industrial societies: a decade of Marxist and critical perspectives', *Current Sociology* 30(1).
Lefebvre, H. (1976) *The Survival of Capitalism*, London: Allison & Busby.
(1979) 'Space, social product and use value', pp. 285–95 in J. Frieberg (ed.) *Critical Sociology*, New York: Irvington.
Lim, L. Y. C. (1978a) 'Multinational Firms and Manufacturing for Export in Less-Developed Countries: The Case of the Electronics Industry in Malaysia and Singapore', Ph.D. Dissertation, University of Michigan.
(1978b) 'Women workers in multinational corporations in developing countries: the case of the electronics industry in Malaysia and Singapore', *Occasional Paper*, Women's Studies Program, University of Michigan.
(1982) 'Capitalism, imperialism and patriarchy: the dilemma of third world women workers in multinational factories', pp. 70–91 in J. Nash and M. P. Fernandez-Kelly (eds) *Women, Men and the International Division of Labor*, Albany: State University of New York Press.
(1983) 'Singapore's success, the myth of the free market economy', *Asian Survey* 23(6): 752–64.
(1987) 'Capital, labour and the state in the internationalisation of high-tech industry: the case of Singapore', in M. Douglass and J. Friedmann (eds) *Transnational Capital and Urbanisation on the Pacific Rim*, Los Angeles: Center for Pacific Rim Studies, University of California.
Lim, L. Y. C. and Pang, E. F. (1982) 'Vertical linkages and multinational enterprises in developing countries', *World Development* 10(7): 585–95.
Lin, T. B. and Ho, Y. P. (1980) *Export-Oriented Growth and Industrial Diversification in Hong Kong*, Hong Kong: Economic Research Centre,

Chinese University of Hong Kong.
Lin, T. B., Mok, V. and Ho. Y. P. (1980) *Manufactured Exports and Employment in Hong Kong*, Hong Kong: Chinese University Press.
Lin, V. (1985) 'Health, women's work and the new international division of labor: women workers in the semiconductor industry in Singapore and Malaysia', Paper for the International Sociological Association Conference on the Urban and Regional Impact of the New International Division of Labour, Hong Kong, August.
(1987) 'Women electronics workers in Southeast Asia: emergence of a working class' pp. 112–135 in J. Henderson and M. Castells (eds) *Global Restructuring and Territorial Development*, London: Sage Publications.
Lipietz, A. (1986) 'New tendencies in the international division of labor: regimes of accumulation and modes of regulation', pp. 16–40 in A. J. Scott and M. Storper (eds) *Production, Work, Territory*, Boston: Allen & Unwin.
(1987) *Mirages and Micracles: the Crises of Global Fordism*, London: Verso.
Locate in Scotland (1983) *The Semiconductor Industry in Scotland*, Glasgow: Scottish Development Agency.
Lojkine, J. (1976) 'Contribution to a Marxist theory of urbanisation', pp. 119–46 in C. G. Pickvance (ed.) *Urban Sociology: Critical Essays*, London: Tavistock.
Luther, H. U. (1978) 'Strikes and the institutionalisation of labour protest: the case of Singapore', *Journal of Contemporary Asia* 8(2): 219–30.
MacInnes, J. and Sproull, A. (1986) 'Union recognition in the electronics industry in Scotland', *Research Report 4*, Centre for Research into Industrial Democracy and Participation, University of Glasgow.
Marcussen, H. S. and Torp, J. E. (1982) *Internationalization of Capital: Prospects for the Third World*, London: Zed Press.
Markusen, A. R. (1985) *Profit Cycles, Oligopoly and Regional Development*, Cambridge, Mass.: MIT Press.
Markusen, A. R., Hall, P. and Glasmeier, A. (1986) *High Tech America*, Boston: Allen & Unwin.
Marx, K. (1967) *Capital, Vol. I*, New York: International Publishers.
Massey, D. (1983) 'Industrial restructuring as class restructuring: production decentralisation and local uniqueness', *Regional Studies* 17(2): 73–89.
(1984) *Spatial Divisions of Labour*, London: Macmillan.
Massey, D. and Meegan, R. (1982) *The Anatomy of Job Loss*, London: Methuen.
McGee, T. G. (1985) 'Joining the global assembly line: Malaysia's role in the international semiconductor industry', Paper to the Workshop on Industrialisation and the Growth of the Labour Force in Malaysia, Canberra: Australian National University, November.
McKenna, S. (1984) 'The Scottish Electronics Industry and Unionisation', B. A. Dissertation, Department of Industrial Relations, University of Strathclyde.
Mingione, E. (1981) *Social Conflict and the City*, Oxford: Basil Blackwell.
Mirza, H. (1986) *Multinationals and the Growth of the Singapore Economy*,

New York: St Martins Press.
Mok, C. H. (1969) 'The Development of the Cotton Spinning and Weaving Industries in Hong Kong, 1946–1966', M.A. Thesis, University of Hong Kong.
Montgomery, D. (1979) *Workers' Control in America*, New York: Cambridge University Press.
Morgan, K. and Sayer, A. (1983) 'The international electronics industry and regional development in Britain', *Working Paper 34*, Urban and Regional Studies, University of Sussex.
Neff, R. (1980) 'Taiwan pushes high technology', *Electronics* 8 May: 100–1.
Newby, H., Bujra, J., Littlewood, P., Rees, G. and Rees, T. L. (eds) (1985) *Restructuring Capital: Recession and Reorganisation in Industrial Society*, London: Macmillan.
Ng, S. H. (1986a) 'Labour' pp. 268–99 in J. Y. S. Cheng (ed.) *Hong Kong in Transition*, Hong Kong: Oxford University Press.
(1986b) 'Electronics technicians in an industrialising economy: some glimpses on the "New Working Class"', *Sociological Review* 34(3): 611–40.
Nichols, T. (ed.) (1980) *Capital and Labour*, Glasgow: Fontana.
Noble, D. (1977) *America by Design*, New York: Oxford University Press.
Norton, R. D. and Rees, J. (1979) 'The product cycle and the spatial decentralization of American manufacturing', *Regional Studies* 13(2): 141–51.
Oakey, R. P. (1984) 'Innovation and regional growth in small high technology firms: evidence from Britain and the USA', *Regional Studies* 18(3): 237–51.
O'Connor, J. (1981) 'The meaning of crisis', *International Journal of Urban and Regional Research* 5(3): 301–29.
(1984) *Accumulation Crisis*, Oxford: Basil Blackwell.
Offe, C. (1975) 'The theory of the capitalist state and the problem of policy formation', pp. 125–44 in L. Lindberg, R. Alford, C. Crouch and C. Offe (eds) *Stress and Contradiction in Modern Capitalism*, Lexington: Lexington Books.
Ohmae, K. (1985) *Triad Power: The Coming Shape of Global Competition*, New York: The Free Press.
Okimoto, D. I, Sugano, T. and Weinstein, F. B. (eds) (1984) *Competitive Edge, The Semiconductor Industry in the US and Japan*, Stanford: Stanford University Press.
Paglaban, E. (1978) 'Philippines: workers in the export industry', *Pacific Research* IX (3 and 4).
Palloix, C. (1977) 'The self-expansion of capital on a world scale', *Review of Radical Political Economics* 9(2): 1–28.
Palma, G. (1981) 'Dependency and development, a critical overview', pp. 20–78 in D. Seers (ed.) *Dependency Theory: A Critical Reassessment*, London: Frances Pinter.
Phelps Brown, E. H. (1971) 'The Hong Kong economy: achievements and prospects', pp. 1–20 in K. Hopkins (ed.) *Hong Kong: The Industrial Colony*, Hong Kong: Oxford University Press.

Phongpaichit, P. (1981) 'Bangkok masseurs, holding up the family sky', *South-East Asia Chronicle* 78: 15–23.

Pollert, A. (1981) *Girls, Wives, Factory Lives*, London: Macmillan.

Portes, A. and Walton, J. (1981) *Labor, Class and the International System*, New York: Academic Press.

Pottier, C. (1987) 'The location of high technology industries in France', pp. 192–222 in M. Breheny and R. McQuaid (eds) *The Development of High Technology Industries: An International Survey*, London: Croom Helm.

Rabushka, A. (1979) *Hong Kong: A Study in Economic Freedom*, Chicago: University of Chicago Press.

Rada, J. (1982) *The Structure and Behaviour of the Semiconductor Industry*, Geneva: United Nations Centre on Transnational Corporations.

Riedel, J. C. (1972) 'The Industrialization of Hong Kong', Ph.D. Dissertation, University of California, Davis.

Roberts, B. (1978) *Cities of Peasants*, London: Edward Arnold.

Salaff, J. (1981) *Working Daughters of Hong Kong*, Cambridge: Cambridge University Press.

Salih, K. and Young, M. L. (1987) 'Social forces, the state and the international division of labour: the case of Malaysia', pp. 168–202 in J. Henderson and M. Castells (eds) *Global Restructuring and Territorial Development*, London: Sage Publications.

Sassen-Koob, S. (1987) 'Issues of core and periphery: labour migration and global restructuring', pp. 60–87 in J. Henderson and M. Castells (eds) *Global Restructuring and Territorial Development*, London: Sage Publications.

Saxenian, A. (1981) 'Silicon chips and spatial structure: the industrial basis of urbanisation in Santa Clara County, California', *Working Paper 345*, Institute of Urban and Regional Development', University of California, Berkeley.

(1983a) 'The genesis of Silicon Valley', *Built Environment* 9(1): 7–17.

(1983b) 'The urban contraditions of Silicon Valley: regional growth and the restructuring of the semiconductor industry', *International Journal of Urban and Regional Research* 7(2): 237–62.

Sayer, A. (1982) 'Explanation in economic geography: abstraction versus generalisation', *Progress in Human Geography* 6(1): 68–88.

(1984) *Method in Social Science*, London: Hutchinson.

(1986a) 'Industrial location on a world scale: the case of the semiconductor industry, pp. 107–23 in A. J. Scott and M. Storper (eds) *Production, Work, Territory*, Boston: Allen & Unwin.

(1986b) 'New developments in manufacturing: the just-in-time system', *Capital and Class* 30: 43–72.

Schiffer, J. (1981) 'The changing post-war pattern of development: the accumulated wisdom of Samir Amin', *World Development* 9(6): 515–37.

(1983) 'Anatomy of a "laissez faire" government: the Hong Kong growth model reconsidered', Mimeo, Centre of Urban Studies and Urban Planning, University of Hong Kong.

Schmitz, H. (1984) 'Industrialisation strategies in less developed countries: some lessons of historical experience, *Journal of Development Studies*

21(1): 1–21.
Scott, A. J. (1980) *The Urban Land Nexus and the State*, London: Pion.
(1982) 'Production system dynamics and metropolitan development', *Annals of the Association of American Geographers* 72: 185–200.
(1983) 'Industrial organization and the logic of intra-metropolitan location: I – theoretical considerations', *Economic Geography* 59(3): 233–50.
(1987) 'The semiconductor industry in Southeast Asia: organisation, location and the international division of labour', *Regional Studies* 21(2): 143–60.
Scott, A. J. and Angel, D. (1986) 'The US semiconductor industry, a locational analysis', *Working Paper*, Department of Geography, University of California, Los Angeles.
Scottish Development Agency (1984a) *North American Companies Manufacturing in Scotland*, Glasgow: Scottish Development Agency.
(1984b) *Electronics Companies in Scotland*, Glasgow: Scottish Development Agency.
Scottish Education and Action for Development (1984) *Electronics and Development: Scotland and Malaysia in the International Electronics Industry*, Edinburgh: Scottish Education and Action for Development.
Seaward, N. (1987) 'Malaysia's electronics industry enjoys a spectacular boom', *Far Eastern Economic Review*, 26 November: 69–70.
Siegel, L. (1980) 'Delicate bonds, the global semiconductor industry' *Pacific Research*, XI(1).
Siegel, L. and Borock, H. (1982) *Background Report on Silicon Valley* (Prepared for the US Commission on Civil Rights), Mountain View: Pacific Studies Center.
Siegel, L. and Markoff, J. (1985) *The High Cost of High Tech: The Dark Side of the Chip*, New York: Harper & Row.
Sit, V. F. S. (1983) *Made in Hong Kong*, Hong Kong: Summerson.
Sit, V. F. S., Wong, S. L. and Kiang, T. S. (1979) *Small Scale Industry in a Laissez Faire Economy: A Hong Kong Case Study*, Hong Kong: Centre of Asian Studies, University of Hong Kong.
Skocpol, T. (1977) 'Wallerstein's world capitalist system: a theoretical and historical critique', *American Journal of Sociology* 82(5): 1075–90.
Slater, D. (1985) *Territory and State in Latin America*, Amsterdam: CEDLA.
Snow, R. (1982) 'The new international division of labor and the US workforce: the case of the electronics industry', pp. 39–67 in J. Nash and M. P. Fernandez-Kelly (eds) *Women, Men and the International Division of Labor*, Albany: State University of New York Press.
So, A. Y. (1986) 'The economic success of Hong Kong: insights from a world-system perspective', *Sociological Perspectives* 29(2): 241–58.
Stigler, G. (1951) 'The division of labor is limited by the extent of the market', *Journal of Political Economy* 59: 185–93.
Sun, L. J. (1983) *The Deep Structure of Chinese Culture*, Hong Kong: Yat Shan Publications (in Chinese).
Sunoo, H. K. (1978) 'Economic development and foreign control in South

Korea', *Journal of Contemporary Asia* 8(3): 322–39.
Szczepanik, E. (1958) *The Economic Growth of Hong Kong*, Oxford: Oxford University Press.
Taylor, J. (1979) *From Modernization to Modes of Production*, London: Macmillan.
Thompson, E. P. (1967) 'Time, work-discipline and industrial capitalism', *Past and Present* 38: 56–97.
Thompson, P. (1983) *The Nature of Work*, London: Macmillan.
Thrift, N. (1987) 'The fixers: the urban geography of international commercial capital', pp. 203–33 in J. Henderson and M. Castells (eds) *Global Restructuring and Territorial Development*, London: Sage Publications.
Tronti, M. (1972) 'Workers and capital', *Telos* 14: 25–62.
Troutman, M. (1980) 'The semiconductor labor market in Silicon Valley: production wages and related issues', Mimeo, Mountain View, Pacific Studies Center.
Tso, T. M. T. (1983) 'Civil Service Unions as a Social Force in Hong Kong', M.Soc.Sc. Dissertation, University of Hong Kong.
Turner, H. A., with Fosh, P. etc. (1980) *The Last Colony: But Whose?* Cambridge: Cambridge University Press.
UNCTC (1986) *Transnational Corporations in the International Semiconductor Industry*, New York: United Nations Centre on Transnational Corporations.
UNIDO (1980) 'Women in the redeployment of manufacturing industry to developing countries', *Working Papers on Structural Change 18*, New York, United Nations Industrial Development Organisation.
United States Department of Commerce (1979) *A Report on the U.S. Semiconductor Industry*, Washington DC: US Government Printing Office.
Vernon, R. (1966) 'International investment and international trade in the product cycle', *Quarterly Journal of Economics* 80: 190–207.
Villegas, E. M. (1983) *Studies in Philippine Political Economy*, Manila: Silangan Publishers.
Walker, J. (1987) 'The electronics industry and other high technology industries in Scotland', in D. McCrone (ed.) *Scottish Government Yearbook 1987*, Edinburgh: Edinburgh University Press.
Wallerstein, I. (1974) *The Modern World-System*, New York: Academic Press.
(1978) 'World-system analysis: theoretical and interpretative issues', pp. 219–35 in B. B. Kaplan (ed.) *Social Change in the Capitalist World Economy*, Beverly Hills: Sage Publications.
(1979) *The Capitalist World Economy*, Cambridge: Cambridge University Press.
(1980) *The Modern World-System II*, Academic Press: New York.
(1983) *Historical Capitalism*, London: Verso.
Walton, J. (ed.) (1985a) *Capital and Labour in the Urbanized World*, London: Sage Publications.
(1985b) 'The third "new" international division of labour', pp. 3–14 in J.

Walton (ed.) *Capital and Labour in the Urbanized World*, London: Sage Publications.
Warren, B. (1980) *Imperialism: Pioneer of Capitalism*, London: Verso.
West, J. (ed.) (1982) *Women, Work and the Labour Market*, London: Routledge & Kegan Paul.
Williams, R. (1980) 'Base and superstructure in Marxist cultural theory', pp. 31–49 in R. Williams *Problems in Materialism and Culture*, London: Verso.
Wilson, R. W., Ashton, P. K. and Egan, T. P. (1980) *Innovation, Competition, and Government Policy in the Semiconductor Industry*, Lexington Mass.: D. C. Heath.
Wong, S. L. (1979) 'Industrial Entrepreneurship and Ethnicity: A Study of the Shanghainese Cotton Spinners in Hong Kong', D.Phil. Thesis, University of Oxford.
Worsley, P. (1980) 'One world or three?: a critique of the world-system theory of Immanuel Wallerstein', *Socialist Register*: 298–338.
(1984) *The Three Worlds*, London: Weidenfeld and Nicolson.
Wright, E. O. (1978) *Class, Crisis and the State*, London: New Left Books.
Young, S. (1984) 'The foreign-owned manufacturing sector', pp. 93–127 in N. Hood and S. Young (eds) *Industry, Policy and the Scottish Economy*, Edinburgh: Edinburgh University Press.
Young, S. and Hood, N. (1984) 'Industrial policy and the Scottish economy', pp. 28–56 in N. Hood and S. Young (eds) *Industry, Policy and the Scottish Economy*, Edinburgh: Edinburgh University Press.
Youngson, A. J. (ed.) (1983) *China and Hong Kong: The Economic Nexus*, Hong Kong: Oxford University Press.

Index

Notes 1. Most references are to the electronics industry and especially semiconductors. 2. Names of firms are indicated by *italics.*